Canadian Women in the Sky

Canadian Women in the Sky

100 Years of Flight

Elizabeth Gillan Muir

DUNDURN
TORONTO

Project Editor: Jennifer McKnight
Copy Editor: Britanie Wilson
Design: Laura Boyle
Cover Design: Sarah Beaudin
Cover Images: Luggage tag: ©MyCreativeLand; Sky: ©diannehope/ MorgueFile. Photo of Vera Strodl Dowling courtesy of Warren Hathaway.
Printer: Webcom

Library and Archives Canada Cataloguing in Publication

Muir, Elizabeth Gillan, 1934-, author
          Canadian women in the sky : 100 years of flight / Elizabeth Gillan Muir.

Includes bibliographical references and index.
Issued in print and electronic formats.
ISBN 978-1-4597-3187-5 (paperback).--ISBN 978-1-4597-3188-2 (pdf).-- ISBN 978-1-4597-3189-9 (epub)

1. Women air pilots--Canada--History. 2. Women in aeronautics-- Canada--History. 3. Aeronautics-- Canada--History. I. Title.

TL539.M87 2015   629.13092271   C2015-905034-0
                                                                                    C2015-905035-9

1   2   3   4   5   19   18   17   16   15

     Canadä

We acknowledge the support of the **Canada Council for the Arts** and the **Ontario Arts Council** for our publishing program. We also acknowledge the financial support of the **Government of Canada** through the **Canada Book Fund** and **Livres Canada Books**, and the **Government of Ontario** through the **Ontario Book Publishing Tax Credit** and the **Ontario Media Development Corporation**.

VISIT US AT

Dundurn.com | @dundurnpress | Facebook.com/dundurnpress | Pinterest.com/dundurnpress

Dundurn
3 Church Street, Suite 500
Toronto, Ontario, Canada
M5E 1M2

Contents

Foreword

I remember reading once that a plane cannot tell if its pilot is a man or a woman. I was profoundly struck by it. *Of course*, I thought to myself, *it's so obvious*. But for most of our history with aviation, women were told their place was not in the cockpit or the flight deck, or anywhere near an airplane — except, perhaps, as a "stewardess," as flight attendants were called back then.

Elizabeth Muir shows through the stories in this book that though they were discouraged from it, women wanted desperately to work in aviation from its earliest days. Through determination, pluck, and training, women like Katherine Stinson Otero, Elsie MacGill, Vi Milstead Warren, Felicity McKendry, Roberta Bondar, Maryse Carmichael, and others that you will meet in the pages of this book, accomplished amazing things and reached their dreams. They flew the first flying contraptions, which were just a collection of struts and wires, all the way through to bush planes, four-engine military bombers, water bombers, and helicopters. They designed and built warplanes and trainers. They went into the skies and into space.

Like many of these women, I never set out to be a pioneer when I became the first female president of the Canadian Aviation Historical Society in its fifty years of existence. I just followed my interests, put my hand up when the opportunity arose, and then worked as hard as I could to do a good job. I had to earn respect, as anyone would, but my gender never caused any commotion other than pleasant surprise among some that a young woman would be interested in aviation history. Even this reaction, however, twigged me to the fact that for some it never occurred to them that women would be interested in this sort of thing — a quiet prejudice all in itself.

Unfortunately there remain those who still vocally resist female flyers, mechanics, astronauts, and other roles traditionally closed to women. Luckily, they are few and far between. When the WestJet passenger wrote his comment about "lady pilots" on a napkin in 2014, it quickly went viral on the Internet. The fact that it happened right around Women in Aviation Day felt particularly ironic. I was so happy to see that critics ranging from journalists in traditional media to everyday people on social media unilaterally slammed it as archaic, sexist thinking, and I hope that gentleman has changed his mind.

The thinking around women's roles in aviation might be shifting, but there needs to be a critical mass in civil, military, and commercial aviation to boost the tiny numbers that currently exist. We are playing catch-up now from the decades when there were hardly any women in the field. It was not that long ago that they had to battle uphill the whole way, conform to very narrow standards, and sometimes even stay invisible to passengers. It has really only been since the 1970s or 1980s when women like Lorna de Blicquy and Rosella Bjornson were making their mark, and perhaps twenty-five years since little girls might encounter more than a handful of examples to follow.

Seeing one pioneer in a pilot's uniform is not enough to push the eager but reluctant dreamers in our lives. Girls need to see female airline CEOs, baggage handlers, ticket agents, and flight crew early and often in the books they read, the shows and movies they watch, and in the media more generally. They need events like Girls Can Fly, Women in Aviation Day, and the Elsie MacGill Foundation's Northern Lights Awards to give them opportunities to experience flight and gain mentors. They need encouragement from parents, teachers, and other adults of both genders in their lives.

You have taken that step by reading this book and perhaps lending or giving it to a girl or woman in your life. Who knows what adventure or career path you have sparked, what passion you have inspired? I have had the opportunity to write about and meet many amazing pioneering women, including some of the ones mentioned in this book, and they are truly inspirational. But many of them are also surprisingly "normal," which makes the dream of working in avionics, aeronautical engineering, or air traffic control seem possible. It makes a young girl think, *hmmm, if she could do it, so can I!*

It may sound strange, but I cannot wait for the day when there are no longer new female trailblazers in aviation. I cannot wait for the day when it is a normal, everyday occurrence to see women in aviation, from bush to boardroom to blasting off into space. Hopefully this book and others like it will help make that happen.

Danielle Metcalfe-Chenail
Historian Laureate for the City of Edmonton, Alberta
Author, *Polar Winds: A Century of Flying the North*

Preface

I do not fly. That is, I do not pilot a plane or a space ship, but I am fascinated by flight. No matter how often the mechanics of flight is explained to me, I consider it one of life's magical mysteries.

I remember how excited I was as a child when I won a short plane ride as a prize for designing a poster for a local fair. As I was helped up, I could hardly believe I was stepping into a plane. My father was watching enviously, for he had never been in a plane either, but I was loath to give up my place. The ride ended far too quickly.

Later, I dreamed of becoming a stewardess, but I was discouraged when I found out you had to train to be a nurse first.

The impetus for writing this small book, however, is somewhat different. Just recently, British Airways told the story of two young children who were visiting the flight deck on one of their planes. The female pilot, Aoife Duggan, asked the children if they would like to fly a plane too. The young boy said yes, but the six-year-old girl replied that she'd like to be an air hostess. "Boys are pilots," she explained to the female pilot.[1]

Events among my friends led me to discover that many young girls are still dreaming of careers that were considered proper for women half a century ago — and not looking at the exciting possibilities of the non-traditional vocations and professions that were closed to women in the early twentieth century, but now are theoretically wide open to all.

I began to look at statistics and discovered to my amazement that many women are still not participating in certain careers, either because they have not been encouraged, or because they do not think them attractive or suitable. Perhaps the most startling example is in aviation. Women make up only six percent of air pilots, flight engineers, and flying instructors in Canada today.[2]

This is the reason for this book — to illustrate the tremendous contribution Canadian women have made to the airline industry over the past one hundred years, and to encourage mothers and grandparents, teachers, and all those who come into contact with young people, to tell the story of women in the air and inspire both girls and boys, women and men, to consider aviation as a career path.

There are so many exciting and interesting stories worth telling, and there are so few pages available. How did I select the women I've included? It was a difficult and somewhat random decision, although all the women included have made an impressive contribution to aviation in Canada and have served as excellent role models. I hope that you will add to this volume by searching out other women who have made a career in this field and listening to their amazing stories.

A few American women are included in these pages — some who inspired Canadian women and some who are an integral part of Canadian aviation history.

Acknowledgements

Many people have contributed their knowledge and expertise towards the completion of this work. A number of the female pilots themselves were able to provide photos and share interesting information: Judy Adamson, Lola Reid Allin, Rosella Bjornson, Roberta Bondar, Dee Brasseur, Judy Cameron, Maryse Carmichael, Adele Fogle, Barbara Ann Scott King (now deceased), Akky Mansikka, Margo McCutcheon, Felicity Bennett McKendry, Mary Ellen Pauli, Julie Payette, and Liz Wieben. Other historians and writers came forward, generously offering photographs and missing data. Among them are: Terry Baker, Nancy Bink, Terry Brunner, Joanna Calder, Will Chabun, Timothy Dubé, Todd Fleet, Renald Fortier, Jim Fortin, Margo Frink, Warren Hathaway, Brian Losito, Dave McKillop, Danielle Metcalfe-Chenail, Hobie Morris, Keith Morris, G.E. Nettleton, James Norton, Rollande Ruston, Marilyn Snedden, James Sulis, Andrew Tom, Jerry Vernon, and Harold E. Wright. Their assistance has been invaluable. I apologize for any omissions.

Unfortunately there is limited information about some of the earliest pioneer women in aviation, but museums and archives have sorted through their collections and found photographs that help to tell the amazing story of courageous and passionate early female "flyers": Canada's Aviation Hall of Fame, Canada Aviation and Space Museum,

Canadian Bushplane Centre, Canadian Flight Museum, Canada Science and Technology Museum, City of Greater Sudbury Heritage Museum, City of Ottawa Archives, City of Vancouver Archives, Comox Air Force Museum, Empire State Aerosciences Museum, International Women's Air and Space Museum, Library and Archives Canada, Mississippi Valley Textile Museum, Musée de l'Air et de l'Espace (Paris), National Defence Headquarters, Skate Canada, Ottawa Public Library, Smithsonian Institute Archives, Toronto Reference Library, and Victoria and Albert Museum Archives.

As always, the Dundurn team were impressive in their specialized assistance, especially my editors, Britanie Wilson, Kathryn Lane, and Jennifer McKnight, and my publicist, Jaclyn Hodsdon.

And Danielle Metcalfe-Chenail and Marc Garneau went beyond the call of duty in writing the foreword and an endorsement for the book. I am extremely grateful to them for taking time to do this in the midst of their daily demands.

And, as with all my writing, my children Deirdre Kathleen Allaert and James Gillan Muir have been steadfast in providing the constant computer help necessary for contemporary authorship, while meeting their family and work obligations.

Introduction

Grace Mackenzie's father ordered her to come home to Toronto at once; she was twenty-two years old and all she'd done was ride in an airplane over New York City, but that was about a hundred years ago when women weren't supposed to be so outrageous as to fly in the sky, even as passengers.

Elizabeth Wieben was hired as a bush pilot to fly passengers into isolated areas in northern Canada. She had to sneak onboard before the passengers, and afterwards, she wasn't allowed to talk over the plane's address system. The airline allowed her to fly their planes as long as the passengers didn't know a woman was the pilot. A lot of people wouldn't fly if they knew that a woman was at the controls. That was only forty years ago.

It wasn't long ago that many people in Canada thought women weren't intelligent or strong enough to fly airplanes, that women were too emotional to be any good in an emergency. Indeed, even early in 2014 a passenger on a WestJet flight left a note on his seat saying that "the cockpit of an airline is no place for a woman" and asked that he be alerted the next time "a fair lady is at the helm" so that he could book another flight.

When planes were in their infancy, a few American women flew in Canada; they were treated as heroes — women such as Alys McKey Bryant and Katherine Stinson Otero. But most Canadians believed that women in this country should stay home, cook and clean, and look after children. Being a pilot, or even flying *in* a plane as a passenger, was "man's work." Women, Rowena Davison explained in 1909 in *An Edwardian Housewife's Companion*, needed to run the home and attend to a husband's wishes and needs, especially after "a [man's] long and arduous day's work," and make "the place called home an abiding joy."[1] As the president of the Regina Flying Club reminded its members in the 1930s, "women's only place in flying is as the mother of pilots."[2]

One recent report on women in the air in Canada explained in an understated fashion, "for some time there was some general social uneasiness about women flying." The report noted, however, that "Canadian women looked to some quite famous 'aviatrixes,' mostly in Britain and the United States, for their inspiration and as role models."[3]

Although flying was not considered suitable for women, many women were determined to fly. They scrimped and saved to pay for flying lessons. Often they were blocked as they tried to carve out a career path in aviation, although some men did support them and help them along the way. Eventually women were accepted as pilots.

This is the story of how a few women in Canada became airplane passengers, then flight attendants and pilots, and finally astronauts. This is the story of their challenges and their successes, their slow march over one hundred years from scandal to acceptance, from the time the first woman climbed onboard a flying machine as a passenger to a female Canadian astronaut's second visit to the International Space Station.

We need to remember that many other women lacked the resources, determination, or strength to overcome obstacles. But the few who were able to persist eventually broke down those barriers so that others could follow. The sky-blue glass ceiling ultimately cracked.

BEFORE THERE
WERE PLANES

1

Up, Up in a Basket

... Through the pathless realms of cloudland
Wishful thousands watched her go
Gazing on the daring aeronaut
As she looked on those below ...[1]

Mme. Godard (?–1877)

On September 8, 1856, the French couple, Eugène (1827–1890) and Mme. Godard, owners of a balloon company who were travelling around the United States, made a gas balloon ascension in Quebec — possibly the first duo to do so in Canada.[2] They recorded two more flights from Montreal on the fifteenth and twenty-second of the same month. Apparently they always went up together. Sixteen years earlier an American, fifty-four-year-old Louis Lauriat, had taken off in the *Star of the East* balloon from Saint John, New Brunswick; however, Mme. Godard appears to be the first woman to go up in Canadian air.

Unfortunately, as the Godards entered Canada, their large balloon, the *Aurora*, was damaged beyond repair; they hastily gathered together an army of fifty seamstresses in Montreal at the future site of Bonsecours Market to stitch a new one to be called *Le Canada*. Mme. Godard had supervised a hundred or more "work girls" in Europe, cutting and sewing varnished calico material to make balloons, each one taking about twelve days to manufacture. Afterwards they added a wicker car, a pigeon cage, a barometer, compass, and thermometer.[3]

Le Canada was inflated with 36,860 cubic feet of gas in the Ste. Anne gas works, and with 20,000 spectators and a band playing, the Godards launched their new balloon from the Wesleyan Methodist Church yard in Griffintown, Montreal. Eugène boasted that their balloons could take up a horse, and to demonstrate, mounted one on a platform under the balloon basket as he rode up into the sky.

Eugène made more than 2,500 balloon flights in France and abroad.

Nellie Thurston, sketched by an American artist, Cliff Tyzik. The balloon bears the name of balloonist and Nellie's second husband and manager, Herman Squire. *Courtesy Hobie Morris.*

Nellie Thurston (1846–1932)

Some years later, in the summer of 1879, "a lady who has acquired a continental fame for her many voyages into space," Miss Nellie Thurston, sometimes described as a Canadian but born in Troy, New York — according to her tombstone in Gravesville Cemetery, New York — awed men and women in the Ottawa Valley by "floating away" in the "monster" balloon *Lorne*, named after the fourth Canadian governor general of that time, the Marquis of Lorne, Sir John Douglas Sutherland Campbell. Launched in the afternoon from McFarlane's Grove in Almonte, Ontario, the balloon was seen by people in Appleton and Carleton Place, and throughout Ramsay and Beckwith townships, sometimes appearing "the size of a tub, then shrunk to the dimensions of a pail."[4]

Friday, August 29 was a "red-letter day" for the small community of Almonte. Bands came from Perth, Brockville, Smiths Falls, Pembroke, Renfrew, Bristol, Arnprior, and Carleton Place. Sporting events included a tug of war between ten men from the fire companies in many of those towns. There was a ten-mile foot race for a forty-dollar purse, horse races, hurdles, and boat races — even a "fat man's race," in which the competitors had to have at least a forty-inch waist and weigh over two-hundred pounds. Trainloads of spectators came from the surrounding countryside. There were not enough passenger cars and boxcars to accommodate all those who wanted to travel. Banners, flags, and evergreens decorating the stores and homes in the town added to the "bustle and confusion" of the day. The *pièce de résistance*, however, was the ascension of the "graceful airship" with its "fair occupant" that reached an "elevation of about three miles." Gate receipts totalled $1,052, much to the relief of the organizers, who had expended about $1,000 for the day.

A newspaper article a few days later noted that watching Nellie in the balloon, "women felt compassion, men expressed admiration, but a feeling akin to awe pervaded all … easy it seemed and yet so daring."[5]

Nellie was in Canada again on September 2, 1891, this time advertised as Nellie Lamont, as part of a fair in Quebec. "The young and pretty and very brave" Miss Nellie Lamont, performed "hot air balloon ascensions and parachute descents" with "Professor" W.W. McEwen, a printer, theatre manager, and smoke balloonist who advertised on postcards.

A six-by-three-foot poster from 1879, advertising the Almonte, Ontario August twenty-ninth "extravaganza," featuring balloonist Nellie Thurston. The poster hangs in the Comox Air Force Museum in British Columbia, part of the Geoffrey Rowe collection of aviation memorabilia. *Photo by Dee Allaert.*

Known variously as Nellie La Mountain, Nellie La Mount, or Nellie La Mont, she was born Ellen Moss and was later adopted by balloonist Ira Thurston. In 1864 she married John La Mountain, "a brave but cantankerous balloonist," then, after a divorce, married Herman D. Squire in 1872.

Nellie went up in balloons over twenty years, mostly throughout the United States at fairs and other public events. She was called "the only female astronaut in America," the first American woman to make solo ascensions in a balloon.[6]

Ballooning was reasonably safe, although serious accidents did happen. Between 1888 and 1914 Canadian newspapers reported at least forty-one fatalities in the United States as a result of ballooning, but none in Canada except for four spectators who were killed. There were, however, a number of non-fatal incidents.[7] For example, in September 1874, as Nellie was descending, her basket got caught in some trees in a remote area and

Fanny Godard (1839–1880) was part of the French Godard family of balloonists, likely a daughter or niece of Mme. and M. Eugène Godard. A celebrity throughout Europe, she also scandalized some with her scanty costume.
© Coll. musée de l'Air et de L'Espace, Paris, France, Le Bourget MA 24672.

she had to climb down a seventy-foot rope. She tried to fasten the anchor of the balloon so it wouldn't take off, but she cut both her hands badly and a gust of wind hurled her to the ground, "the anchor catching in my clothing, tearing through with lightning force [and] I barely escaped its entering my side … my heart sank within me, my balloon gone, my clothing torn to shreds, my hands badly lacerated and covered with blood, my body bruised and I in a wet cold uninhabited wilderness, I knew not where."

The French balloonist Fanny Godard (1839–1880) broke an arm in 1879 when the wind prevented her balloon from landing for several hours, and somehow she and her companion both suffered injuries.[8] Howard Squire crashed into the Wesleyan Methodist Church in 1874 in Brockville, Ontario, after the mesh and strings of his balloon tangled in the church steeple. There had been a strong wind that day. His balloon ascent was the feature of the Brockville Dominion Day celebrations and, with about 10,000 people watching, he narrowly escaped death. After a dramatic rescue by the local inhabitants, but badly injured, he gave up ballooning and became Nellie's manager.[9]

In London, Ontario, two other women aeronauts ran into trouble. In September 1892 a Miss Mandell fell onto a roof while she was making a parachute descent with her dog Aerio. Not to be deterred, she tried again three days later, and that time fell onto a "steam threshing engine." In 1894 a Mlle. De Vean fell twelve feet to the ground after she collided with a house chimney at a "Western Fair."[10]

In spite of such incidents, Nellie continued her career for some years, going up in a balloon at least 140 times. Ballooning or "balloonmania" was fashionable and drew huge crowds.[11]

However, a Canadian would not be a passenger in a plane until 1909.

THE EARLY YEARS

2

Daredevil Female Passengers

Dolena MacKay MacLeod (c.1887–1969)

Dolena MacKay MacLeod was about twenty-two when she climbed into an airplane at Baddeck, Nova Scotia, the first known Canadian woman to fly in a plane.[1]

Little information is available about Dolena and what possessed her to attempt such a risky venture. Her family was terrified: her husband, Jim, their two small children, and several relatives, including her cousins the MacRae sisters, watched in fear. Many of these early "fragile, unstable flying machines" crashed.[2]

It was September 1909, a few months after the first flight in Canada, only six years after the Wright brothers first flew their primitive contraptions in the United States. The *Baddeck* flying machine was named after the village in Nova Scotia where Dolena flew.[3]

Seventeen-year-old engineer and athlete Frederick W. "Casey" Baldwin was the pilot, the first Canadian to fly a plane.[4] He owned the Canadian Aerodrome Company that had made the *Baddeck*.[5] Baldwin had been working with Alexander Graham Bell, the inventor of the telephone; JA Douglas McCurdy, the son of Alexander's secretary and later governor general of Nova Scotia; and Mabel Bell, Alexander's wife, whose organizational skills assured the success of the venture.[6]

Mabel Hubbard Bell, Alexander Graham Bell, and their two daughters, Elsie and Daisy, 1885. Although deaf from childhood scarlet fever, Mabel was the organizing force behind Canada's early airplanes.
Photo by Dr. Gilbert H. Grosvenor. Image ID 243090, National Geographic Society.

Even with such a "qualified" pilot, however, Dolena's family was anxious. It must have looked to them as if she were flying on a kite. Thankfully she landed safely.

One hundred years later, Julie Payette would be celebrated as the second Canadian woman in space, and the first to enter the International Space Station. Before that, however, women struggled against societal norms just to become passengers on a plane.

Grace Mackenzie (1888–1946)

The wealthy Toronto socialite, Grace Mackenzie, the second Canadian woman to fly in a plane, was much more daring than Dolena. In 1910, Grace went to New York City with her two sisters to watch the dark-haired, dark-eyed, tall French pilot Count Jacques Benjamin Marie de Lesseps (1883–1927) fly in an air show.

Jacques had a fascinating background. He was the second last child in a family of nineteen children. His father, Count Ferdinand de Lesseps, a celebrated engineer who built the Suez Canal and began work on the Panama Canal, was about seventy-eight when Jacques was born.

Jacques had been the second pilot to fly across the English Channel in 1909, claiming the *Daily Mail*'s Ruinart prize for this thirty-five-minute feat, even though he had to land in England without the power of his fifty-horsepower engine. Jacques became an international hero, invited to air shows throughout Europe and North America. Wherever he went, women swooned; it was said that his eyes were like those of a basset hound.[7]

The next year Jacques made a forty-nine-minute flight over Montreal, the first flight over a Canadian city. The Montreal Royal Automobile Club of Canada had offered him $10,000 plus $5,000 for expenses — an enormous amount at that time. A 10,000-seat grandstand was built, and people paid from fifty cents to two dollars to attend. Women cried "Vive le compte!" The Mohawk elders at the near-by Caughnawaga reserve made him an honorary member.[8]

Later that month, Jacques became the star of Toronto's first aviation meet. In his Blériot monoplane, *La Scarabée*, he left the small field north of Toronto, owned by the mining magnate W.G. Tretheway, circled three times, then flew south to Toronto on a loop of twenty miles in twenty-eight minutes. Almost everyone in the city craned their necks to watch this strange sight in the sky; a few who weren't aware of the event were terrified. After Jacques flew to 3,500 feet at seventy miles per hour, Tretheway gave him a $500 prize for his courage.

The next day, a writer exclaimed, "What the future of it is we know not, but as we watched last night's performance in awe-struck and reverent amazement, the vision of the future held us in its spell."[9]

Grace and her family were evidently awe-struck, too. Grace's well-known father, the entrepreneur Sir William Mackenzie, owner of the Toronto Street Railway and president of the Canadian Northern Railway, met Jacques at a private club, invited him to the Mackenzie home "Benvenuto" for dinner the next day, and offered him a tour of the Trent Canal on the family yacht, *Wawinet*, before Jacques left for New York City.

Later, when their father was away on business, Grace and her sisters, Kathleen and Ethel, followed Jacques to New York. He flew in the air show there, where he won the Great Aerial Tournament at Belmont, also

Two of the Mackenzie sisters flying with Compte Jacques de Lesseps in a Blériot aircraft. Grace may be the one facing towards the front.
The Frank Coffyn Collection, Empire State Aeroscience Museum.

flying around the Statue of Liberty, which had been given to the United States by his father in 1884. Jacques took Grace up for a five-mile flight; afterwards, he gave her sisters a short ride, too.[10] The newspapers printed stories about their flights.

When Grace's father read about the plane rides in New York City, he was furious. He may have been impressed by Jacques, but it was outrageous for his daughters to fly with him. He sent a telegram ordering the girls to come back to Toronto at once.

Jacques followed Grace back home. Evidently she reconciled with her father, for she married the dashing flying ace the next year in January, in what appears to have been a lavish wedding at St. James Church in Spanish Place in London, England, with the reception at Claridge's. The best man was one of Jacques's brothers-in-law, an engineer, the 11th Count of Mora, Marquis de Ruvigny. The guest list included other nobility and well-known celebrities.

Grace wore a gown of white chiffon embroidered with pearls, with a long satin train, and carried an enormous bouquet of orchids and lilies-of-the-valley. Her three bridesmaids — her sister Ethel and possibly

Grace Mackenzie (Comptesse de Lesseps) on her wedding day, January 26, 1911, to Compte Jacques de Lesseps (left male) at St. James Church, Spanish Place, with the reception at Claridge's Hotel, London. The bridesmaids are Ethel Mackenzie, Mable Meagher, and one unidentified, possibly Grace's other sister. The best man is the Marquis de Ruvigny, Count of Mora.
Photo by Lafayette Ltd., London. Courtesy Victoria and Albert Museum Archives.

her sister Kathleen, as well as a relative, Mable Meagher — were in pink chiffon with mauve sashes and hats of pink tulle trimmed with skunk, each carrying a large bouquet.[11] Skunk was evidently the fashion of the day, for both Grace's mother's outfit and Grace's going-away dress were also trimmed with it.

After the wedding, Jacques and Grace moved to Paris, where Jacques flew airplanes during the First World War, defending the city against German Zeppelin attacks. France awarded him the Croix de Guerre and the Legion of Honour: by 1918, he had completed ninety-five combat missions. Meanwhile, Grace helped out in a hospital.

After the war, Jacques and Grace returned to Canada with their family of two boys and two girls. Elisabeth Hélène "Tauni" de Lesseps

(1915–2001), one of their daughters, became a well-known sculptor and painter, with artwork exhibited around the world and with three bronzes in the White House. Their two sons tragically died young.[12]

As the director of exploration and chief pilot for La Companie Aerienne Française, Jacques made aerial maps of the Gaspé coast. "What a wonderful vision also provides the colour contrast between the vast peninsula and the sea which surrounded on one side, velvet forest green and silky, artistically draped according to the whim of the mountains and valleys on the other, the blue sky, united to the Gulf under the dazzling sky," he wrote, impressed with the Canadian countryside.[13]

In 1927, Jacques's plane went down during a storm while he was mapping the area. He and his flight engineer, Theodor Chichenko, were killed. An impressive monument was erected in 1932 for both men in the town of Gaspé, with the inscription in part:

> A LA MEMOIRE DE JACQUES DE LESSEPS, CHEVALIER DE LA LEGION D'HONNEUR, CROIX DE GUERRE, DISTINGUISHED SERVICE CROSS U.S.A. NE A PARIS EN 1883. DEUXIEME TRAVERSEE DE LA MANCHE EN AVION 1910. LE PREMIER SURVOLA MONTREAL ET TORONTO 1914–1918. QUATRE FOIS CITE A L'ORDRE DU JOUR. PERDU EN MER 18 OCT 1927 AU COURS DU LEVER AERIEN DE LA CARTE DE LA GASPESIE. CITE A L'ORDRE DE LA NATION FRANCAISE. SON CORPS REPOSE DANS LE CIMETIERE DE GASPE …[14]

Olive Stark (c.1877–1927)

The third Canadian woman to fly as a passenger, Olive Stark, had a narrow escape in 1912; the plane she was in almost crashed.[15]

Olive's husband, William M. "Billy" Stark, a well-known Canadian auto racer, had attended the Curtiss Aviation School in San Diego, California, owned by the famous aviator Glenn Curtiss, to learn how to fly. Billy came back to British Columbia with his pilot's licence (#110) and a new $5,500 *Curtiss Flyer* biplane constructed of wire, bamboo, and spruce wood, and covered with fabric. It was noted that earlier he had driven the first "gasoline-propelled vehicle on the streets of Vancouver" in 1901, and was fascinated by automobiles and engines.[16]

In 1912, he flew in a Vancouver air show near Minoru Park and took Jim Hewitt, the *Daily Province's* sports editor, up for a flight. Hewitt wrote that a trip in an elevator in a "modern" skyscraper gave one more thrills than a journey with Billy Stark, the ascent was made so evenly:

> Once in the air, the strong rush of the wind kept me so busy hanging on to the rigging to prevent being swept out of the machine backwards that I had no occasion to worry about falling. My only fear was that I might be blown out into space. We dashed through the air at a pace which approximated, as the aviator informed (me) afterwards, about forty miles an hour and the feeling was somewhat similar to that which would be experienced on the cowcatcher of a locomotive travelling at top speed against a head-on gale …[17]

Billy asked Olive to go up with him next. There was only one proper seat, but she was tiny enough to sit on a board Billy had latched to the lower wing, just to the left of his seat, as Hewitt had done. Also like

Billy Stark at the wheel of his airplane with his wife, Olive. While there are two wheels on the plane, it is thought that Olive never piloted it.
City of Vancouver Archives.

Hewitt, her feet dangled in the air and she had to hang on to wires; there was no seat belt or harness. Olive was bundled up for the flight, wearing a pair of boy's pants, which at the time was considered almost as scandalous as riding in a plane. Olive was said to have looked "both proud and apprehensive." Evidently she had been a passenger on a number of earlier flights in the United States.[18]

During Olive and Billy's forty-mile-per-hour flight, Olive's heavy woollen tam flew off her head and got caught on a propeller blade. This caused the plane to shake and almost tip over, but the tam dropped off and the plane steadied itself. When the brief flight ended, Olive was so cold and tired from hanging on to the wires that she had to be lifted down.

Afterwards, Olive flew often with her husband; his later planes had two seats. He continued flying in exhibitions around British Columbia. In the fall of 1915 British Columbia's first Aero Club began and Billy's flying machine was bought by the club to use as a trainer. Billy was hired as an instructor.[19]

3

The Flying Schoolgirl: Katherine Stinson

Katherine Stinson Otero (c.1891–1977) was a flying superstar in Canada, although she was an early American pilot.

It is not known exactly when Katherine was born, but it was sometime between 1891and 1899. She pretended to be younger than she really was. With long brown curly hair, a tiny figure, a mischievous smile, and a charming Southern accent, she seemed young. She weighed only eighty pounds and wore girls' clothes, rather than pilots' outfits. She was fragile and dainty, definitely feminine, and was described as a brunette version of Mary Pickford, the Canadian-born movie star. She was known as "The Flying Schoolgirl."

Katherine had wanted to study music in Europe, but that was expensive. She discovered that flying at fairs was as quick a way as any to make money, so she learned to fly. Then she found that she liked flying so much she gave up the idea of a career in music.

Katherine began flying as a teenager or in her early twenties (her date of birth is uncertain). She was the oldest of four children — Katherine, Marjorie, Jack, and Eddie — and the fourth woman in North America to hold a pilot's licence.[1]

During the First World War, private flying was banned, as airplanes and fuel were desperately needed for the war effort. In the United States

Sisters Marjorie C. Stinson (left) and Katherine Stinson just after flying in Katherine's biplane, 1913.
Smithsonian National Air and Space Museum, 2007-5474.

only Katherine was allowed to fly because she raised money for the Red Cross. Her plane had a large red cross on its tail, visible for many miles. In 1917 she made a 610-mile flight that took nine hours and forty minutes, raising $2,000,000 from pledges, approximately $38,000,000 in comparable currency today.[2]

Other family members were also pilots, all of them working at Stinson Aviation Company, the flying school owned and operated by Katherine and her mother in Texas, where they trained young men to fly — many of them young Canadian pilots going to war. Katherine's younger sister Marjorie was the chief flying instructor. The company also rented out and sold airplanes.[3]

Katherine brought her single-seat yellow biplane to Canada for the first time in 1916, flying at exhibitions and fairs across Canada — in Calgary, Edmonton, Regina, Winnipeg, and Brandon, Manitoba. At Brandon she was made a princess by Chief Waukessa of the Sioux tribe, "as tribute to the admiration of his people." Because of the war, she had to have written permission from the military authorities to fly in Canada.[4]

A lover of thrill, she made daring nosedives that were almost straight down, then she'd fly straight back up again. She became the first woman to loop-the-loop in Canada.[5] She attached flares or Roman candles to the back of the wings of her plane, and while she was flying she released the flares, producing an early form of skywriting.[6]

Katherine was the first person to fly alone at night; flares helped people see her. Once, at 10:30 p.m. in Winnipeg, she asked people who owned cars to turn on their headlights so she could see where to land. Her flight was in aid of the Manitoba Patriotic Fund, but it had been delayed because a strong wind storm had damaged her plane.

Katherine's most famous flights in Canada were in 1918. That fall in Peterborough, when it was raining, she was asked if she were afraid to fly. "I am never afraid," she answered. "Why should I be?" Observers reported that even in the rain she did "quick swoops downward" and then "a quick climb with the machine banked at a dangerous angle."[7]

Katherine believed that women were less afraid to fly than men. "At least, more women than men have asked to go up with me. And when I took them up, they seemed to enjoy it," she explained.[8]

Katherine also flew that year in Toronto, Sherbrooke, Quebec, and Ottawa, but she earned a place in Canadian history for making the first cross-country flight in Alberta — the first airmail flight in the Canadian west. On July 9, wearing a heavy coat and carrying a horseshoe for good luck, Katherine flew from Calgary to Edmonton delivering a mailbag containing 259 letters.[9] Unfortunately her plane had trouble and she had to make a brief emergency stop at Beddington near Red Deer. A fuel leak stained some of the letters. But it was a record 200-mile flight in two hours.[10] All along the route, the telegraph tapped out her progress to the crowds waiting in Edmonton, and the Edmonton–Calgary train raced her between Lacombe and Morningside. The *Edmonton Journal* reported the next day that "Flying as true as an arrow the bird-like figure hove into sight from the south, and it was only a few minutes before the whirring of the propeller could be easily heard. Flying at a great height,

Miss Stinson gracefully circled the grounds, coming down by easy stages until in a favourable position to land against the wind."[11]

Katherine made her last visit to Edmonton in 1959 to open the Edmonton Fair.[12]

Katherine's plane was always transported in pieces and then put together where she was flying. Called the *Katherine Stinson Special*, it was made up of all kinds of spare parts, as it was difficult to get new ones during the war. Her plane was described as a junk shop hybrid. One special feature was a small vanity table, a compartment with a lid that opened into a mirrored table, where she could touch up her makeup before she appeared in public.

A replica of Katherine's plane was made for the Air Museum in Edmonton, requiring fifteen people volunteering 20,000 hours over four years to create it. This was first displayed on July 9, 2006, as part of a celebration of her airmail flight. Edmonton pilot Audrey Kahovec, dressed in a costume similar to the one Katherine had worn, reenacted the famous flight. "I was excited to be part of this historical event," Audrey said. She had earned her pilot's licence as a teenager in the Royal Air Cadets. Also, to mark the occasion, Jim Bruce, a well-known aviation artist, painted the Stinson plane in flight as it followed the CPR train tracks.

Katherine's love of speed was not exclusive to airplanes. In 1918 she established a world record for driving a Fiat car at the Edmonton Fair, travelling one mile in one minute and fifteen seconds — faster than cars are allowed to go on most city streets today.[13]

Katherine became ill in 1920; she'd contracted tuberculosis while working as an ambulance driver in France during the war, and had to stop flying. She married airman Miguel Antonio Otero Jr. and worked as an architect in New Mexico.

4
It's a Bird ... It's a Plane ... It's Madge Graham

A giant bird-like flying machine brushed the tops of the tall trees. Thousands of people were waiting on the ground below. "Here it comes," they cried, as they caught a glimpse of *La Vigilance* through the mist and the clouds. Many of the men and women had never seen a flying machine before, but all of them were curious and excited at the news that a woman was onboard.

Madge Vermeren Graham (1882–1983) had told the local newspaper that she'd be her husband Stuart's navigator from Nova Scotia to Quebec, her very first trip in a plane. What would such a daring woman be like? Flying in 1919 was dangerous. Besides, women in those days were supposed to stay home to cook and clean and look after children, not fly in the sky.

Born in Belgium of English parents, "Madge" as she was called (although her real name was Marguerite Marie), was not a typical Canadian woman of her time. She was beautiful and tiny; she was an artist; she loved birds; and she loved to ride horses. She was adventurous and capable. Once when she was riding alone in India, where she lived for a few years, three men tried to pull her off her horse and kidnap her. Madge fought them off with her riding crop, the horse bolted, and Madge hung on, escaping her would-be captors.

Madge met Stuart in England while he was serving in the First World War.[1] He was wounded during the devastating battle of Ypres in 1915, and they crossed paths while he was convalescing in a hospital in Ramsgate, England. When he recuperated he transferred to the Royal Navy Air Service and they married.

Now, as the Curtiss HS-2L plywood flying boat landed on the water, the crowds cheered as if Madge were a rock star. People poured out of shops, houses, and offices and ran back and forth between wharves, not knowing where the plane would tie up.

One of the more touching scenes was a reunion between Madge and an old friend, Mrs. E. M. Archibald. Madge had sent her a telegram saying, "Arriving St. John by seaplane, 2 or 3 o'clock. Try come and see me by boat." No one in St. John had ever received such a telegram!

The press went crazy. Newspaper reporters took pictures and interviewed Madge. Much of the resulting *Daily Telegraph*'s extensive coverage, "ARRIVES HERE BY BIRD ROUTE FROM HALIFAX," focused on her.

When a reporter asked her if she felt safe, she replied, "With my husband at the wheel, I would go" anywhere, but "when newspaper people confront me, I'm all nerves which are never permissible aboard an aircraft."[2] The *Daily Telegraph* reporter asked why she had come on this hazardous trip, and she answered, "I have come for my husband's sake, but also because it will give other people confidence."[3]

Described as "a charming English girl," Madge told one reporter, "You have a lovely country," and she wished everyone could see "our lands as the birds see them." Obviously impressed by her, the reporter noted in his newspaper story that day that "the so-called stronger sex had got to go into active training if it hopes to keep pace with the weaker in the new things that are bound to turn this old world upside down."[4]

Madge's 800-mile trip that June took almost five days, a trip that today would take a little over one hour, but she and her American-born husband Stuart had to stop frequently to refuel the plane and when the weather was bad.[5] It was an uncomfortable trip. There were only basic instruments on the plane: a compass, an air and wind speed indicator, and a turn and bank indicator. Some of the time, the plane was waterlogged. It was so noisy that they couldn't hear each other talk; Madge rigged up a six-foot-long clothesline so that they could send messages back and forth across the distance that separated them in the plane.

After they landed in St. John, New Brunswick, Mayor White gave them the keys to the city and free tickets to *Il Trovatore*, the opera that was playing locally that night. Madge was the only one who could hear the music because she'd worn earplugs during the flight. Stuart and the mechanic onboard were temporarily deaf and, according to Madge, "could only watch the singers' mouth opening and closing."[6]

On the flight all three had to sit in open cockpits. An elasticized canvas cover that Madge had made to fit around the opening where she sat — in what was once the gunner's cockpit — covered all of her except her head. But even with this, she got soaked whenever they took off and landed on water without a windshield to protect her. Her seat was at the very front of the seventy-four-foot-wide plane.[7] When they landed at Rivière du Loup a large wave went over the front of the plane; Madge completely disappeared under the water for a moment. Stuart was quite upset, as Madge had never learned to swim, but she wasn't fazed.[8]

All three fliers were warmly dressed because it was windy and extremely cold up high at around 700 feet where they flew. Madge wore a heavy leather Kapok-lined coat, helmet, and goggles. Actually, they had hoped to fly even higher, but there was too much fog. There were no aerial maps at that time, so Madge had to watch for lakes and rivers to figure out where they were.

The three-person crew repeated the trip that same month with a second plane. Both planes were to be used later in the area for air surveillance, mainly to look for bush fires and to make maps of the area.[9] Stuart became the first Canadian bush pilot, and later, in 1920, Madge flew with him to a remote part of the woods in northern Quebec where she cooked meals for mining surveyors.[10]

Madge and *La Vigilance* received a mixed reception wherever they landed. Some men and women thought Madge was scandalous. One woman asked her why she was wearing men's "britches." On the street before a flight another woman demanded of Madge, "Where are you going in that outfit?" Madge replied, "I'm going flying. What would you wear?"

On other occasions, envious and supportive women presented Madge with flowers and chocolates. In some isolated areas people were quite frightened by the apparition of the monstrous plane and its begoggled occupants: native Indians were convinced this was a manifestation of *Machi-Manitou*, the evil spirit; some French Canadians called the plane *l'échelle volant*, the flying ladder, because of the appearance of the criss-crossed wires.

When the American explorer Admiral Richard Byrd heard about Madge's flight, he said that "flying seaplanes over land is suicidal and taking a woman along is criminal," obviously believing that women should not be that adventurous or should at least be more protected.[11]

Madge never learned how to fly. After her flights in 1919 and 1920 she stayed home with five children — three from a previous marriage and two from her marriage to Stuart, "leaving the flying to her husband."[12] Later Madge wrote, "As I look back on that summer (1919) fifty-two years ago, I think I must have been addle-pated. It was a stupid, fantastic thing to do."[13]

Now in the National Aviation Museum in Ottawa, *La Vigilance* crashed on takeoff in 1922 and stayed at the bottom of a lake until 1969, when it was salvaged and painstakingly restored.

Madge Graham at Rockliffe, Ottawa, with the remains of HS-2L *La Vigilance*, 1970. *CAVM 15430*.

5

The American Influence

The First Flight

On December 17, 1903, on a sandy hill in Kitty Hawk, North Carolina, the American Orville Wright (1871–1948) flew an airplane for the first time to approximately 119 feet. The flight lasted twelve seconds.[1] That same day his brother, Wilbur Wright (1867–1912), flew fifty-nine seconds, even further than Orville at 853 feet.

With canvas-covered wings, wooden propellers, and a wooden frame, the plane had cost $1,000 to build, the equivalent of about $25,000 today. The pilot had to lie flat on the plane's body.

Earlier Orville and Wilbur Wright had built bicycles, but all machines fascinated them and they had decided they wanted to build one that would fly. The two youngest brothers in a family of seven, they grew up watching their mother make toys for them, fixing anything that was broken in their home.

Orville and Wilbur were reasonably well off and had time to tinker with machines. Determined to fly, they experimented with different kinds of flying machines, building and rebuilding and testing them over and over again to get the right design. They even used a small homebuilt wind tunnel to test wings and propellers. Their first planes were more like

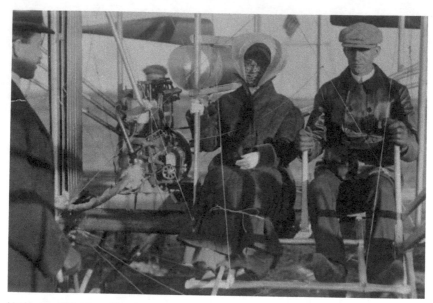

Katharine Wright and her brother Wilbur Wright, February 15, 1905, on her first flight, Pau, France.
MS-1, Wright Brothers Collection 18-4-7, Wright State University.

gliders than airplanes, and they concentrated on developing pilot control, contrary to their competitors who tended to focus on building powerful engines. When asked why they were successful, Orville answered that they never gave up; they continually tried and tested new designs themselves.

The Wright brothers demonstrated flying all over the world. Their younger sister Katharine (1874–1929) — a high school teacher who later married an old sweetheart, newspaperman, and math teacher, Henry Haskell — flew with them in 1909 in France. As Wilbur noted, "If ever the world thinks of us in connection with aviation, it must remember our sister."[2] In 1909 *The World Magazine* reported in a story about Katharine, "She is as fearless if not as ingenious as her brothers. Europe has been watching and praising her as few American women have been watched and praised."[3]

Katharine (Kate) had been the mainstay in the Wright's home after their mother died. She acted as a hostess for her father, a bishop in the Church of the United Brethren, and looked after her brothers' correspondence. Later, in Europe, she became Orville and Wilbur's social guide, meeting easily with royalty and other prominent persons, making up for skills her brothers lacked. It is no wonder that they asked her to accompany them in Europe. Rumours of her many talents, such as sewing wing coverings and designing planes — rumours she always denied — circulated in France especially, to the point that she was awarded the Legion of Honour along with Wilbur and Orville. It appears, however, that she was more interested in looking after her brothers' interests than in flying itself.

Harriet Quimby (1875-1912)

The dark-eyed American Harriet Quimby, a movie screenwriter, photographer, theatre critic, actress, and prominent newspaperwoman, was the first American woman to get her pilot's licence, in 1911. Receiving licence #37, she became the second licensed woman in the world. "The men flyers have given out the impression that aeroplaning [sic] is very perilous work, something that an ordinary mortal should not dream of attempting. But when I saw how easily the male flyers manipulated their machines, I said I could fly," she wrote.

Harriet should have been credited as the first woman to fly in the United States as well, but a woman called Blanche (Betty) Stuart Scott (1885–1970) "accidentally" became the first female American pilot in 1910 when the plane that she was allowed to taxi mysteriously became airborne! But Blanche was not given a licence at that time. She was taking flying lessons, however, and later became an accomplished stunt pilot.

A colourful woman with a sense of style, Harriet wore a purple satin costume with a hood when she flew, instead of the men's clothing that other women adopted. She flew across the English Channel in 1912, three years after Jacques de Lesseps, making the flight in fifty-nine minutes. Before her flight, people told her that she would never make it. In fact one male pilot offered to dress in her clothes and fly in her stead, changing clothes afterwards to make it look as if she had flown. This incensed her and "... made me more determined than ever to succeed," she said.[4] The papers carried very little news about her achievement, however. The day before, the ocean liner *Titanic* had sunk, and that consumed the interest of the public and the press.

Later that same year Harriet plunged to her death flying in the Third Annual Boston Aviation Meet. She had been known as a particularly safe pilot who wrote about pre-flight checks and the necessity of wearing a seat belt. "Flying is a fine, dignified sport for women, and there is no need to be afraid so long as one is careful," she wrote.[5] It is thought that a gust of wind or mechanical failure caused her plane to lurch, and she and her passenger were thrown from the plane. The irony is that they were not wearing seat belts. Flying in the air at the same time, watching Harriet's plane, was Blanche Scott, by then a qualified pilot.

Harriet not only enjoyed the speed of flight, but some time earlier she had written an article on the excitement of being in a motor car when the chauffeur suddenly sped up to seventy miles an hour, even though at that time cars were designed to reach a hundred and over. "You are now going at about seventy miles an hour and you feel the swift currents of air produced by the mad flight of the machine ... and the car careens virtually on one wheel, and the whole machine seems lifted up in the air and you come down to earth again with a jump...."[6]

Later Harriet was a passenger in an automobile race with the first woman to drive a car 3,600 miles across the United States, of which only 152 miles were paved.

The White Flyer

The American "Tiny" Alys McKey Bryant (1880–1954), called "The White Flyer" because she wore large white bloomer-style pants when she rode a bicycle, was the first woman to fly a plane in Canada.[7]

Alys was teaching home economics in California when she read a 1912 newspaper ad: "WANTED: YOUNG LADY TO LEARN TO FLY FOR EXHIBITION PURPOSES."[8] She answered the ad, attracted by adventure — she was athletic and rode around on a motorcycle. She was hired, so she built her own plane from a Curtiss-type pusher biplane with one engine and two seats, which had been wrecked. She taught herself to fly, but also took lessons from John Milton "Johnny" Bryant, a flying instructor whom she later married. She made a name for herself in the United States in California, Oregon, and Washington as a daring "aviatrix," flying at fairs and other exhibitions. The public was fascinated by flight, especially that of female pilots.[9]

In July 1913 Alys flew in Vancouver on three different days, having been asked to fly as part of the entertainment for the Prince of Wales and his young brother, the Duke of York, King George V's sons who were visiting Canada that year.[10] She held the altitude record for women in the United States, flying in Seattle at 3,100 feet; in British Columbia she also established a record, flying at 2,200 feet.[11] She flew at Victoria "in the worst flying conditions she had ever been up in."

Johnny was also flying in the gusty wind conditions, but unfortunately his flying machine crashed onto the roof of a building in Victoria, and he became Canada's first aircraft fatality, the only pre-war flying victim in Canada. It was late afternoon on the third day. Johnny had been in the air only five minutes, about eight hundred feet above the city, when those on the ground saw the plane dive until it was only two hundred feet above ground. The right wing collapsed, and the plane crashed.[12] It was Johnny's six hundredth flight in the same aircraft. It was thought that an earlier passenger had strained the seat and some of the wires had not been fully repaired. Johnny and Alys had been married for only ten weeks.

American pilot Alys McKey Bryant, March 1913.
Trans N38, City of Vancouver Archives.

It is not clear what Alys did afterwards. Some reports claim that she stopped flying; others vary, saying that she flew for movie shoots, set more altitude records for women, trained U.S. pilots during the First World War, worked at an aviation plant, helped build the first airport at Washington, D.C. by clearing trees and brush, and that she became a deep-sea diver. Perhaps all hold some truth.[13]

"My one thought is that WINGS may be used … NOT for destruction, but for making more friendly and understanding relationships between the nations of the earth," she wrote.[14]

Amelia Earhart (1897–c.1937)

Amelia Earhart wasn't the first American woman to fly, but she was the most famous. When she was young she kept a scrapbook about other women's accomplishments, and even took an auto repair course. She and her sister climbed trees, hunted rats with a rifle, and generally lived as tomboys. They kept worms, moths, katydids, and a tree toad.

Amelia went to Toronto, Ontario, around 1917 to visit her sister Muriel who was studying there, and remained to train as a nurse's aide, joined the Canadian Red Cross, and worked at Spadina Military Hospital. There she prepared food for the wounded men and handed out medicine, but during the Spanish flu pandemic she became a patient herself.

It was in Toronto that Amelia fell in love with flying. Often going to watch the planes at Armour Heights Airfield north of Toronto, and to the Canadian National Exhibition in 1919, she became fascinated by planes and the young Royal Flying Corps pilots' skills.

Amelia returned to the United States, and two years later had her first flight over Los Angeles. "As soon as I left the ground, I knew I myself had to fly," she wrote. She worked as a telephone company clerk and photographer to pay for her lessons, and later as a social worker in Boston.[15]

In 1932 she flew nonstop across the Atlantic Ocean, the first woman to fly that route. She fought fatigue, a leaky fuel tank, a cracked manifold that spewed out flames on the side of the engine, and ice on the wings. That same year she flew non-stop across the United States, about 2,408 miles, determined to do all the things that men had tried. "Girls are shielded and somewhat helped so much that they lose initiative and begin to believe the signs 'Girls don't' and 'Girls can't' which mark their paths," she exclaimed.[16]

A celebrity, Amelia was sought after as a lecturer and writer, and to provide paid-for endorsements, also designing a line of functional women's clothing and lightweight luggage. She realized she could make her living from aviation. While she wore heavy gear to protect herself from fumes and the weather when she was flying, she wore attractive blouses, scarves, and makeup when photographers were around to show other women that "flying was safe and restful."[17]

Her last flight was in 1937 when she, her co-pilot Jim Noonan, and her plane disappeared over the Pacific Ocean near Howland Island, trying to circumnavigate the globe. "Adventure is worthwhile in itself," she wrote. She was last seen on takeoff from New Guinea on July 20, having completed 22,000 miles. To this day no one knows what happened, although there has been much speculation as to their fate.[18] Some say the plane ran out of fuel, crashed into the water, and sank. Others suggest that she landed on some other island, possibly the South Pacific Island of Nikumaroro, where bone fragments and remnants of artifacts, such as pieces of a cosmetic jar, have been found. And of course there are the conspiracy theorists who claim that she was a spy for the U.S. government, landing on one of the Japanese-occupied islands where she was taken prisoner, or that she in fact returned to the U.S. and lived in obscurity into old age!

Amelia wrote of an earlier trip, "Obviously I faced the possibility of not returning … once faced and settled there really wasn't any good reason to refer to it."

DREAMS CAN
COME TRUE

6

Early Pioneers

Young women in Canada emulated Amelia Earhart. Just as she kept scrap-books about other women celebrities, women kept scrapbooks about her.

Scores of women became so obsessed about flying that they were called air-minded or air-crazy, constantly dreaming about flying with the birds and the clouds, so much so that the cover of a 1931 issue of *Maclean's* magazine featured an attractive young woman dressed in helmet and breeches in front of an airplane propeller, looking longingly to the sky. The year before there had been a cover of a woman in helmet and goggles with a male pilot going off on their honeymoon with tin cans and other material tied to the tail of their plane.[1]

But it wasn't easy for Canadian women to learn how to fly. First they had to find someone who would teach them. Most had to work hard to earn enough money to pay for lessons; then when they learned how to fly they had a difficult time finding a company that would hire them as pilots.

One event that inadvertently helped women was the government's funding of local flying clubs, which encouraged female pilots. In fact it was around this time that the male president of the Regina Flying Club said that, "Woman's only place in flying is as the mother of a pilot."

In 1929 *Saturday Night* summed up the prevalent Canadian attitude: "A considered opinion among nerve specialists is that, while woman is quite equal to managing a car or a plane in normal conditions, and will probably use more care or common sense, when it comes to meeting the unexpected she is far more likely to lose her head."[2] Even much later on, in the 1990's, one of the female pilots pointed out that there was still a large old boys' club atmosphere in the realm of professional pilots.

There were, however, a number of Canadian women who fought these attitudes and roadblocks, who discovered that dreams could come true for them. These are the stories of a few of these pioneer women.

Eileen Vollick Hopkin (1908–1968)

From her bedroom window teenaged Eileen Vollick could see everything that was going on at a local aerodrome, "the cutting down of trees, the dumping of load after load of cinders." Each day as she drove her car past the construction site a small voice whispered to her, "Go ahead, brave the lion in his den and make known your proposition to him."[3] A petite beauty queen, Eileen had always wanted to fly. She felt that "opportunity" was calling to her in a thousand ways; she felt the urge to fly, to become a pioneer and "blaze the trail" for women in Canada.[4]

"I have never been afraid to go after anything I wanted," she said, so one day she "braved the lion." She went to the Jack Elliot Flying School in Hamilton, Ontario, and, despite the fear that he might laugh at her, asked Jack if girls could learn to fly. She was seventeen, the age when men learned, but the owner wasn't so sure about a woman. He didn't laugh, but he made her wait until she was nineteen; the government changed the rules after they heard that a woman had applied.[5]

During Eileen's first flying lesson the instructor tried to frighten her with loops and spins, but to his surprise she only loved it and responded, "Can we do it again?" Eileen wasn't at all afraid, having already made several parachute jumps from 2,800 feet from the wing of a plane into Burlington Bay the summer before.[6]

Eileen took her flying lessons in the winter when the temperature was below zero, wearing a fur-lined flying suit because the cockpit was open to the weather. She was only five feet one inch tall; she needed cushions to see out the window of the airplane cockpit. Her lessons were at 6:00 a.m., before she went to work at 8:30.

Eileen's mother evidently supported her venture, for she altered one of Eileen's father's mechanic's suits so that Eileen could look like the rest of the "boys" in the classroom. This would not have been considered as radical attire as when the early women passengers flew almost two decades earlier, for in 1933 the *Bracebridge Gazette* reported that "The latest thing in men's clothes … will be women."

Eileen's sister Audrey notes that Eileen was full of spirit and "everybody loved her … She could play the piano. She loved parties, when we would all gather round and sing."[7]

On March 13, 1928, after only sixteen hours of instruction on a ski-equipped "Jenny" plane, Eileen received pilot's certificate #77; she was the first Canadian woman to become a pilot. She had to fly 175 miles cross-country for her test. The Hamilton Cotton Company, where she worked as a textile analyst and assistant designer, had given her time off to take the federal aviation examination.

Eileen Vollick in one of Jack Elliot's planes just after she passed her test to receive her flying licence — the first woman in Canada to do so.
Canada Aviation and Space Museum.

Pilot Eileen Vollick Hopkin.
Canada Aviation and Space Museum, 21543.

"As the winged monster roared and soared skyward," she wrote, "the pure joy of it gave me a thrill that is known only to the airman who wings his way among the fleecy clouds.... Blasé people say there is no thrill in flying ... but there is for me. This is not the thrill of danger, because there is really very little danger nowadays. It is more the thrill of the unusual, the novel. Besides all that there is a beauty up there that people on the ground have no idea about," she remarked after her examination.[8]

There were a lot of stories about her flight. Jack Elliot had a nose for publicity. He called the newspapers and they came and took pictures of Eileen at the controls. But considering the public's insatiable appetite for anything to do with flying at the time, especially if it involved women, the stories of Eileen's incredible accomplishment seem scanty. *The Hamilton Spectator* (March 14, 1928) devoted five paragraphs and a photo to her and the other ten new male pilots in the middle of the paper, with the heading "Hamilton Girl Earns Right to Pilot Airplane;" *The Toronto Daily Star* (March 15, 1928) at least had her on the front page but in a small paragraph in the bottom left-hand corner, titled "Hamilton

Girl First Canadian Woman Pilot;" and *The (Toronto) Evening Telegram* (March 15, 1928) tucked her into page twenty-seven with an even smaller paragraph, entitled "Gets Pilot's Licence."

Just the day before the *Telegram* had devoted much of its front page and a number of inside columns to the English Captain, Walter Hinchcliffe, and his failed flight across the Atlantic with passenger Honorable Elsie Mackay, Lord Inchcape's daughter; the *Star* (March 1, 1928) had a large photo of Margaret Butler who was flying airmail in the United States; and the *Spectator* (February 28, 1928) carried a front page story of Mildred Johnson of the American Air Transport Association, who had been flying since 1921, among other aviation stories, all about non-Canadians.

Eileen loved adventure and found it in many ways. Later that year she became an aerobatic pilot and skydiver. Numerous companies asked her to demonstrate their airplanes, but she turned them down. Amelia Earhart invited her to go on a flying tour with her and another pilot, the famous American racing star, Jacqueline Cochran, born Bessie Lee Pittman.[9] But in 1929 Eileen and James Hopkin married and moved to New York City. There they raised two daughters, Joyce (Miles) and Eileen (Barnes.)

A Canadian postage stamp showing Eileen in a Curtis Jenny plane was issued in 2008 to celebrate her achievements. A terminal at Wiarton Airport — the town in which she was born — was named after her; it was the first Canadian airport to name a terminal after a woman.[10]

Enid Norquay McDonald (1912–2006)

Enid Norquay's father was a magistrate for the Northwest Territories, her mother an avid curler and celebrated horsewoman.[11] Her two grandfathers were also well known — one a former premier of Alberta and the other the first mayor of Edmonton.[12]

Enid, however, wasn't interested in politics; her only real passion was flying. She was keen on horseback riding and excelled in all sports, but she was especially fascinated by everything that flew — butterflies, bees, hummingbirds, and planes. She went as often as she could to the Edmonton airport and stared longingly at the first mail-carrying planes, begging to ride on one.

Finally Enid's parents let her go on a flight to Winnipeg. She had a cold, but she felt like a bird, and was so happy that she ate a whole box of chocolates on the first part of the trip. As a result she was so sick that she threw up, using all eight paper bags on the plane. The pilot removed her from the plane at Moose Jaw, and it was then and there that Enid decided to learn to fly, "to show the boys that girls were not sissies."[13]

When she was old enough to take lessons, she had to leave home at 5:00 a.m. to go to the airport. She received pilot's licence #928 in 1931 — the first female pilot in Alberta, the third in Canada. That year she visited England, where she realized another dream — meeting Amelia Earhart. She even flew Amelia's plane over England.

After Enid married the next year she stopped flying because her husband disapproved, but she never lost her interest in things that flew. She looked after orphaned hummingbirds, which became so tame that they would perch on her fingers. Enid turned her energies and talents to fundraising for charities, writing newspaper articles, and giving lectures. She took part in racquet sports, swimming, curling, and golf, winning many awards.

Louise Mitchell Jenkins (1890–1986)

As the first Canadian woman to buy her own plane, Louise Jenkins's kids thought she was great. Unfortunately their opinion was contrary to some of the women on Prince Edward Island who felt Louise was disgraceful.

A beautiful, tall, slender, wealthy brunette from "an affluent family in Pittsburgh, Pennsylvania," Louise was an avid horseback rider who wrote poetry and music. She became a vegetarian after visiting a meat-packing plant, and worked as a nurse and ambulance driver in England during the First World War; she skied in Switzerland and rode camels in Egypt.

On Armistice Day in 1918, in London, England, Louise married Jack S. Jenkins, a prominent Prince Edward Island doctor whom she had met during the war. Both "well-connected," their best man was the Canadian business tycoon, Lord Beaverbrook, Max Aitken.[14]

"Dr. Jack," as everybody called him, had his own planes, a runway, and an airplane hangar on a large picturesque estate of almost three hundred acres, with beautiful gardens sloping down to the North

River. Jack had offered the land to the city of Charlottetown for an airport, but they didn't take up the offer, so he built Upton Airport himself. He removed trees, fences, and stone hedges until there was enough landing space. Then he built an aircraft hangar capable of housing three small planes. The airport was licensed in 1932, and Jack leased it to Canadian Airways until 1938.

Jack's inspiration for the airport was a visit from Erroll Boyd in 1930, just before Erroll made a historic Atlantic crossing, becoming the first Canadian to fly across the North Atlantic and the first of any nationality to make the flight outside the summer season.[15] Erroll had been grounded by foul weather in Prince Edward Island for ten days, enjoying the Jenkins's hospitality. Then he left for England, landing on the Scilly Isles.[16]

Jack's family raised cattle and show horses, and trained racing horses. Louise had the only Rolls Royce in the province, but with four children to look after (one of the children, a son, died in early childhood), she found life in the country lacking in stimulation.[17] Louise also appears to have been influenced by Erroll Boyd, learning to fly soon after his visit, but she was also offered airplane rides by other pilots who landed on their farm, and enjoyed them so much that she was determined to fly herself. In 1932 she bought her own red and silver plane. She wanted to publicize her province by putting PEI on the plane's registration, but she had to go all the way to Ottawa to Prime Minister R.B. Bennett, a friend of Lord Beaverbrook, to get permission. The newspapers called Louise "The Daring Lady Flyer."

Described as charming but also strong-willed, Louise went to Quebec to learn to fly; some neighbours thought it was appalling that she would "neglect" her children for that reason. She also took lessons in Florida, where she and the children spent much of the winter. She earned her pilot's licence early in 1932 with Madge Graham's husband, Stuart, as her flight examiner. In February that year Louise made a record non-stop flight from Montreal to Charlottetown in four hours and forty minutes.

Louise was also considered "airplane crazy." She flew almost every day wearing a leather flying suit. "I flew to add excitement to my life," she

Pilot Louise Mitchell Jenkins standing proudly beside her red and white plane. *Royal Aviation Museum of Western Canada, Winnipeg, Manitoba.*

explained.[18] She flew her son Jackie to boarding school and dropped off his clothes, landing on the schoolyard. She flew clean laundry to her daughter Joan (born 1920) at summer camp. She took ill people to the hospital, delivered medical supplies and mail, all by plane. She and her husband Jack staged a number of air shows at the farm for the public's entertainment.

Their daughter Jessica (born 1921) caught the flying "bug" as well. She'd often ride her pony to the airport to try and find someone to take her up in an airplane for a ride. At age ten she fell madly in love with her mother's flying instructor, Gath [sic] K. Edwards, later a pilot for Trans-Canada Airlines.[19]

When Louise's husband became sick in 1933, her friends persuaded her to stop flying in case something happened to him and she became a single parent. Jack, however, lived for many years, and for a while afterwards every time Louise heard an airplane it is said that she cried.

After Jack died in 1972 and her children and grandchildren had grown up, Louise returned to the United States to live in Connecticut.

Marion Alice Powell Orr (c.1918–1995)

By 1932 more than twenty women across Canada had earned a pilot's licence, but most women couldn't afford to take flying lessons, let alone buy a plane. One such woman was shy, rebellious Marion Powell Orr, the youngest of four girls.

When Marion was small she would swing for hours on her family's two-seater swing, pretending she was on a flying machine. She used to jump off roofs in "contraptions she made" herself. She walked six miles to an airport to watch planes flying. By the time she was fourteen, she was totally obsessed with flight.[20]

Marion's parents both died when she was young, and when she was fifteen and just out of grade eight Marion and a sister left home to work at a factory. Marion was determined to fly. She went without new clothes and makeup, she stopped eating lunch, and she walked to work instead of taking the bus. "Flying came first," she explained.[21] She earned about ten dollars a week and had to save for six years before she could afford flying lessons at six dollars an hour. She could pay for only fifteen minutes at a time.

Marion received her private pilot's licence on January 5, 1940, when she was twenty-two. Even though the plane's engine stopped during her test flight, she managed to land safely. The next year she had her commercial licence.[22]

Marion had a variety of jobs. She was hired as the manager and chief flight instructor at St. Catharine's Flying Club, becoming the first woman in Canada to operate such a club. She worked as an aircraft inspector at de Havilland Aircraft, qualified as an air-traffic control assistant, and was hired as a control tower operator in Goderich, Ontario.

When the war effort closed all the flying schools and private airports not training military personnel, she tried to enlist, but the RCAF (Royal Canadian Air Force) turned her down flat. It was then that she found out that British Overseas Airways was hiring women to ferry airplanes in the British Isles, and she was able to join that operation (see chapter 12).

After the war Marion managed to find jobs instructing and flying charters, and then in 1949 she bought the bankrupt Aero Activities

Pilot Marion Alice Powell Orr.
*Canada Aviation and Space
Museum, 30956.*

Ltd. in Toronto. The bank wouldn't lend her money for this venture, but she had several friends who helped her out. She was the first Canadian woman to own a flying school and an airport.[23]

Marion built her own small office, brick by brick. She and her friends used pick axes and shovels to dig ditches and take stumps and stones out of the ground for a new runway in Maple, Ontario, where she was forced to relocate. She had to visit Prime Minister Louis St-Laurent personally in Ottawa to get permission to build the airfield, as many of the local people were opposed to having an airport there. However, an influential local dentist, Dr. Phillip Macfarlane, had thought the land would make a great airfield and urged the community to support Marion's initiative. The airport opened in September 1954 and at its peak handled 2,500 to 3,000 flights a month. Marion was in charge of seven instructors and owned ten airplanes. There were two grass runways.[24]

Later, in 1961, Marion became the first Canadian woman to fly a helicopter, and flew for the Ontario Provincial Police. One day, while instructing a student on a helicopter, the engine failed. The machine crashed and Marion broke her back, although she recovered enough to keep instructing pilots for a short while.

During her varied career Marion logged 23,000 hours flying time — more than 17,000 hours as an instructor. That would be the same as flying all day and night every day for more than two-and-a-half years. During fifty years as an instructor Marion taught 5,000 men and women to fly.

In 1993 she was awarded the Order of Canada.

Vera Strodl Dowling (1918–2014)

Vera watched seagulls where she was born in England, the youngest girl in a family of five. She dreamed of flying, and she envied the seagulls, watching them for hours on end. She even tried jumping off a roof with an umbrella, luckily suffering nothing more than a cut lip and a broken nose.[25] She experimented with balloons, but that didn't work either. She built airplanes out of apple boxes and sacking. Curiously, while she wanted to fly, she couldn't stand the feeling of dropping suddenly on swings, diving boards, and roller coasters.[26]

At a school outing on the Isle of Wight, Vera had her first airplane ride. She had to borrow five shillings from Samuel Church, the headmaster, for the ride, and it took her six months of work in her dad's nurseries to pay it back. As she got out of the plane that day, she said to the pilot, "One of these days, I'm going to fly one of these myself."[27] Five years later she was the chief flying instructor at that same airfield.

Vera's parents were Danish, living in England, and when Vera was fourteen her mother took the children from England to Denmark. There Vera got a job translating for a shipping company, but when she mixed up starboard and port (right and left) and the ship turned onto a sandbar and keeled over, she was fired.

When she was sixteen Vera left Denmark to go back to England on her own to learn to fly. She scrubbed floors, washed dishes, and waitressed in a restaurant to earn enough money to pay for lessons. She worked six-and-a-half days a week, but could afford only a twenty-minute lesson every two weeks. However, she received her pilot's licence when she was nineteen, and the next year earned a glider's licence.

Vera found a job as an inspector in an aircraft factory at twenty-five shillings a week, barely enough to live on raw carrots and milk. She was constantly sick from the paint fumes in the factory. Then she found a job as a test pilot, sometimes working thirty-six-hour shifts.

During the Second World War, she flew with the Air Transport Auxiliary (ATA), recording many adventures and narrow escapes. Once she landed a plane in weather so bad that she had to wait for the sky to clear before she could taxi to the control tower. The operator was flabbergasted. "We don't understand you women. Five of our chaps with all the radio, radar and all the stuff on board and with an entire crew wouldn't fly in this, and you come in … lovely landing. HOW DO YOU DO IT?"[28] Another time it took her almost three hours to get off the runway at Prestwick at Glasgow; the grassy runway was a sea of mud.[29]

"Though we women had to work extraordinarily hard, we found that we were as capable as anyone else at the business of flying and we were accepted as equal among pilots. To me it seems that everyone accepts good performance," Vera said.[30]

Pilot Vera Strodl Dowling. Her jacket, bearing the names of all the planes she flew, is now in a museum in Denmark.
Courtesy Warren Hathaway.

After her wartime experience Vera found a job flying for a company in Sweden. She was known as "The English Woman Pilot," the only female pilot flying commercially in that country at that time, and even considered "the most experienced and accomplished pilot in Sweden." As a result she received a great deal of publicity in newspapers, becoming too stressed to continue.[31]

Vera came to Canada in 1952, answering an ad for a flying instructor in Alberta, making her the first female flight instructor in that province. She thought of Canada as a land of milk and honey. "I couldn't get over the luxury of everything," she said. She worked so hard that she bought her own plane in the early 1960s.[32]

Once when she was flying as a commercial passenger from Edmonton to Europe, her male seatmate wondered what the airplane sounds were. "That's just the trimming of the aircraft as we circle," she replied.

"How do you know?" he asked.

"I taught the pilot how to fly," she said.

The man thought she was "putting him on," but when the pilot came and asked Vera to go into the cockpit with him, her seatmate was impressed.[33]

Vera piloted 150 different kinds of airplanes and flew 30,000 hours. "That translates into well over three million miles in the air," she calculated, "a distance roughly equivalent to seven round trips to the moon."[34]

"To Vera," a friend wrote, "flying an airplane is as natural as breathing, eating, sleeping, walking...." But she was terrified of riding in a car.[35] On her eighty-sixth birthday Vera parachuted with a friend.[36]

Vera relied heavily on her religious faith, pointing out that "as a pilot, you realize what a speck you are and that you can't do anything alone ... Perhaps my experiences ... will encourage others to reach out for a goal that seems to them at first to be beyond their grasp."

Felicity Bennet McKendry (born 1929)

When Felicity was young, she sent fifteen cents and two Quaker Oats box tops to the cereal company in exchange for a Captain Sparks "How to Fly" kit.[37] She had been given an allowance of twenty-five cents a week for completing certain chores around the house and the barn, and she used her money to buy the kit.[38] She received a cardboard mock-up of an airplane instrument panel with dials that could be moved, a control column, throttle, and rudder pedals, and a booklet called "How to Fly." Then, after school, she'd run home to tune in to the radio to listen to a three-minute flying lesson with sound effects, using the Captain Sparks cockpit.[39]

Felicity grew up on a farm in Spencerville, Ontario. When she drove the tractor, she pretended she was flying. Choosing a reference point on the horizon, she practised keeping straight lines for "take-off" and "landing."[40] She watched training planes fly over their farm, and wanted to be in one. She made balsa wood model airplanes, launching them with elastic bands, and one year she won first prize for her flying model in the model aircraft section at the Spencerville Fair.[41]

Felicity read everything she could about airplanes, including articles in *Mechanics Illustrated*. She remembers finding a fifteen-cent comic-styled book, *The Romance of Flying*, which was "thirty-six pages filled with thrilling stories of pilots … aces of World War I and heroes of World War II," but the only woman in it was Amelia Earhart.

Felicity, however, found encouragement. One day she "sidled up to her teacher's desk at recess," a shy ten-year-old girl, and shared her dream of learning to fly and becoming a pilot. Instead of "scoffing" at her, the teacher said, "Some day you will."[42]

Felicity raised chickens on the farm. When she was twelve, she exchanged an "oven-ready" rooster with her Aunt Helen for her first airplane ride. The Rotary Club had provided chickens to worthy applicants, and forty-six egg-laying hens provided Felicity with a steady income at the time.[43]

She constantly dreamed of being a pilot and even tried to get an aviation job, but at that time there was no place in a cockpit for a woman. "I don't know why I became so interested in flying. It must be the wonder and the mystery of it, the principal of flight, the freedom you feel once you are up there," she says.[44]

Her backup plan was to become a stewardess, but she grew too tall — almost five-foot ten-inches. She decided to teach instead. It was while she was teaching at the Ontario School for the Deaf that she finally earned her private pilot's licence at the Kingston Flying Club, using her salary to pay for her lessons. She decided flying was what she really wanted to do. "I was enamoured of flying," she said. "It just awed me."

In 1951, in a letter she wrote to her sister Connie, she exclaimed, "... Enough of that! Now for the really thrilling news ... I soloed on Sunday, April 29! Yip, I SOLOed! ... I don't remember being at all nervous.... It's absolutely wonderful up there by yourself..."[45] The next year Felicity placed first for the Kingston Flying Club in the Webster Trophy competition, St. Lawrence Zone trials. "That's like winning the Stanley Cup of Aviation," she says.[46]

By 1953 Felicity was one of only seven female instructors in Canada, out of more than four hundred. She had quit her teaching job, and when September rolled around she was relieved not to have to go back to school — the first time since the age of six that she hadn't returned to a classroom in September. "What a sense of freedom," she wrote.[47]

The Kingston Flying Club, where she worked, provided "airmail" when the local Wolfe Island Ferry was frozen in, in the winter. "At a pre-arranged time," Felicity remembers, "we would deliver the [mail] satchel, usually flying one of the Club's high wing Fleet 80 Canucks.... You would establish the approach, slide the window back to project your arm and satchel out and at the exact time release it (the satchel). The target was on the ice ..."[48]

Pilot Felicity Bennet (McKendry) on her first day instructing flying, April 30, 1953. *George Lily*, Kingston Whig-Standard.

One of Felicity's students, Spence McKendry, put his name down for a lesson whenever she was available; he would later become her husband in 1955. They often flew together. After Felicity and Spence met they discovered that they had both sent away for the Captain Sparks flying kit when they were young.

Felicity and Spence had two children, and after their children were in school Felicity taught flying again. One of her students was especially notable — astronaut Marc Garneau, whom she taught to fly in 1986.

Felicity taught flying in planes on wheels, floats, and skis. She's still in awe of flight, but she explains it easily: "It's all in the lift. Take a small square of bathroom tissue between your thumb and first finger. This represents the curve of an aerofoil. While you blow over the top of this square, hold your other hand to prevent the air going below the wing. What happens? The paper lifts."[49]

"It's been a magical ride, honestly," she remarks. "Receiving the Elsie MacGill Northern Lights Award in the pioneer division was so VERY special, and unexpected."[50]

Like other pilots, Felicity is active in the community and the volunteer arena. She's received the Dr. Morton Shulman Award for her work for the Parkinson Society.[51] It has been said that "her calm and gentle manner, combined with her skill as a pilot, made her one of the best-loved and respected instructors in the area." Over her career she taught 1,100 Canadians to fly.[52]

A postage stamp, introduced May 1, 2013, celebrating the sixtieth anniversary of Felicity becoming a flight instructor, shows her standing by an airplane strut.

Dee Brasseur (born 1953)

Canadian Deanna (Dee) Brasseur, one of the world's first two female fighter pilots, says, "Women need a total belief in themselves because there's nothing they can't do if they decide to do it."[53] Deanna was one of the first three Canadian women to receive military wings for operational duty in 1980. Now she works as a motivational speaker.

When she was young she'd watch the airplanes come in for a landing at the airport where her family was stationed when her father was in the military. "I remember thinking specifically to myself, 'Boy, they're really lucky they get to do that because if girls were allowed to, I would like to do that,'" she remembers. "Then in seventeen years I went from typist to fighter pilot," she says.

Dee decided to become part of the military when she was nineteen, serving as a dental typist in the orderly room in Winnipeg. Then a trial program for women called SWINTER (Study of Women in Non-Traditional Environments and Roles) helped her advance. In 1981 she became one of the first three women to graduate with their wings, and the first flight instructor at Moose Jaw, Saskatchewan. She also became the first female flight commander out of Cold Lake, Alberta, and she and Jane Foster were the first two women to qualify to fly fighter jets.[54] "Flying," Dee said, "presented the ultimate physical and mental challenge, definitely the job for me, and I was getting the chance to be a military pilot."

Deanna tells young girls who approach her to "work very hard at school — to study and get good marks — and if they're interested in a flying career, they need to be committed and not stop at any point along the way.... In other words," Dee says, "I just tell them to go for it."

Over the years all of these outstanding female pilots discovered that dreams can come true, with imagination, determination, and hard work. As Canadian pilot Carrie Bennett (born 1974) says, "Your dreams will only take you as far as you allow them to; determination and persistence are the keys to success."[55]

7

Barbara Ann Scott: Queen of the Blades

Not all early women were passionate about a career in aviation. Some of them, like Barbara Ann Scott (1928–2012), whose passion was figure skating, learned to fly just for fun.

Eleven years old when she won her first Canadian national figure skating championship, Barbara Ann Scott King went on to win two European championship titles, a World Figure Skating title, and an Olympic gold medal in 1948.

Tiny blue-eyed Barbara Ann worked hard, training up to eight hours a day on skating rinks. But she still found time to enjoy other activities such as horseback riding. Then, one summer, she learned to fly.[1]

In the mid-1940s Barbara Ann's boyfriend and his pals spent the summer learning how to fly a plane. The boys thought they were better than the girls, so Barbara Ann took flying lessons to keep up with the boys.[2]

She had a habit of bouncing the plane when she landed. One time she kept the plane door open on her side to see if she could avoid it and instead she landed in a mud puddle, covering her brand new red woollen coat with mud.[3]

In 1947 the president of Trans-Canada Airlines made Barbara Ann an honorary captain of TCA's Skyliners, the airline's employees, proud that such a worldwide celebrity was also a pilot.[4]

Olympic skater Barbara Ann Scott as a pilot, refuelling her plane.
Skate Canada Archives.

8

In the Captain's Seat: Rosella Bjornson

Rosella Bjornson (born 1947)

The guidance counsellor at Rosella's high school laughed when she told him she wanted to be a pilot. He said girls couldn't be pilots, but she became the first Canadian female pilot for a major commercial airline. Later, in 1990, she would advance even further to become the first female captain on a scheduled airline, Canadian Airlines International.[1]

When she was a child in Alberta, Rosella sat on her father's knee in their family plane. "My father taught me all he could without me actually at the controls,"[2] she recalls. Her playhouse was an abandoned warplane without wings — an "Anson Mark II" — where she imagined she was flying. Her dolls and her sisters were the passengers.

By the time she was sixteen Rosella had already flown for several hours with her father. When she was seventeen her parents gave her formal flying lessons at the Lethbridge Flying Club as her birthday present. She received her licence in a record two months. Rosella studied geography and geology at university, but her first love was flying. While she was at university she helped organize the University of Calgary Flying Club.

Six feet tall, with blue eyes and light brown hair, Rosella became the first female pilot for Transair in 1973, the fourth largest airline in

Canada. There were 2,799 male pilots in the Canadian Pilots Association, and Rosella — the newspapers nicknamed her "Transair's Newest Bird."[3] Rosella had applied the year before, but her father told her "Don't get your hopes up, girl." In fact, she had been turned down by all the major airlines, and took a job as a flight instructor at the Winnipeg Flying Club instead with hopes of finding out where the jobs were.[4]

On her first flight on Transair on July 13 she made the pilot's announcement over the plane's public address system. One of the businessmen on the plane said to a stewardess, "A woman jet pilot! Somebody's joking." When they landed he went to the cockpit to see for himself. Many of the passengers were delighted and asked for an autograph.

But Rosella found it difficult at times, although her quiet manner and skill earned her respect from her colleagues. The airline had difficulty with her two pregnancies — there was no provision in the rulebooks for that!

Pilot Rosella Bjornson (Pratt) wearing her captain's uniform with four stripes on her epaulettes. Once, a cleaning lady in the women's washroom saw Rosella in her pilot's uniform and tried to chase her out, thinking it must be a man.
Courtesy Rosella Bjornson.

After she married Bill Pratt, he sometimes flew as her co-pilot. If he were doing the cabin announcements he would say, "Ladies and gentlemen, on behalf of my wife the captain …"[5]

Rosella often volunteered with youth programs. In 1990 she was featured in an Alberta government poster campaign, "Dream/Dare/Do," to encourage children to set goals and work towards them. She speaks of the need for female pilots to remain focused and determined if they are to succeed.[6]

Rosella was inducted into the Canadian Aviation Hall of Fame in 1997. She retired from Air Canada in 2004. Ten years later the Canadian Ninety-Nines — an organization of women pilots — issued a postage stamp featuring her as a newly hired Transair First Officer and a Canadian Airlines captain.[7]

"As a captain of a Boeing 737 with Canadian Airlines, I feel that I have reached the goal I set for myself when I was a young girl flying with my father in his light aircraft," she says. Now she can devote her time to her other special interests: her home and her children, Ken and Valerie.[8]

DURING THE
SECOND WORLD WAR

9

Angels in the Clouds: Early Stewardesses

Are you:
Under 30 years of age?
Unmarried?
Between 5'2" and 5'6"?
Under 125 lbs?
Then perhaps you can qualify as an air stewardess![1]

When Pat Eccleston Maxwell (born c.1916) saw this newspaper ad, one of her nursing pals dared her to apply.

A pilot friend suggested that Lucile Garner Dennison Grant (1910–2013) apply, too. Lucile and Pat were lucky: they were chosen as the first two Canadian stewardesses on Trans-Canada Airlines. More than a thousand women had applied. Neither Lucile nor Pat had ever been to an airport before, let alone flown on an airplane.

"I felt as if I'd been given a million dollars," Pat said, when she learned she'd been accepted.[2] For Lucile it was "an adventure, everything was new about it; it was a brand new life for women."

In 1938 Trans-Canada decided to hire stewardesses, thinking that if young attractive women were flying on airplanes, other people wouldn't be afraid and would fly, too. "If this wisp of a girl could do it, then certainly the businessman could."[3]

Lucile was sent to Seattle in the United States to United Airlines "to learn how things were done." That airline already had stewardesses on its flights. Later Pat was put in charge of new recruits.

Lucile designed their first uniform — a neat blouse, a grey two-button jacket, skirt, and a "pill box" hat made in wool gabardine by Vancouver's foremost tailor. TCA felt that a flight attendant had to "remind passengers of the girl next door rather than a *femme fatale*," so these were "serious" business suits, but they were not supposed to resemble the pilots' outfits. Soon, however, Lucile redesigned the uniform because nobody liked the grey. They liked the new uniform: a navy blue suit, white blouse or a crewneck sweater, and white gloves.[4]

In the late 1930s and early 1940s the most exciting job for Canadian women was an airline stewardess. "Stews," as they were called, travelled to far-away cities, they met movie stars, executives, and other interesting people. They were considered glamorous, even though their parents had to give written permission for them to be hired, and they earned a lot of money. Pat had been earning $75 a month as a nurse. Her salary increased immediately to $125 a month as a stewardess.

The first two Trans-Canada Airlines stewardesses, Lucile Grant (left) and Pat Eccleston (right), July 1938.
Courtesy Canada Aviation and Space Museum, AC-LGC02.

By March 1939 Rose Crispin Lothian and Lela Finlay McKay had also been hired as stewardesses and Lucile had set up the first stewardess school in Winnipeg, with fifteen women enrolled. Stewardesses became so popular that the airline had hired twenty-eight by 1940.[5]

But the working conditions were difficult. If a stewardess put on weight, got married, or turned thirty, she had to resign. She couldn't wear glasses. She had to pretend to be happy even when she wasn't. She had to learn about weather and aircraft design and anything else a passenger might want to know. She couldn't be tall because the aisle ceilings in the planes were low. Pat was five-feet, three-and-a-half inches and she had to bend down to walk up the aisle.

When a stewardess went to work she saw demanding posters like this one:

Stewardess:
Is your:
Smile friendly and sincere?
Posture erect and poised?
Hair short and styled?
Makeup neat and natural?
Uniform clean?
Blouse fresh and pressed?
Nails manicured and polished?
Shoes repaired and shined?
Hose seams straight?[6]

Lois McDonald Whalley noted: "We had to wear white gloves in the summer and navy blue gloves in the winter. We had to wear girdles, we had to wear nail polish, and we had to wear hats." Only certain colours of nail polish and lipstick were acceptable, but later on there was one concession in the summer: the women could wear leg paint instead of nylons.[7] More than one-third of all the stewardesses quit every year because of such restrictions. Sometimes they found the work was too demanding. Often, however, they left because they married someone they met on the plane.[8]

At first the airline hired only nurses because passengers often became sick. The newspapers called the women "ministering angels." One news reporter declared that they had to "combine the comeliness of Venus with the capabilities of Florence Nightingale." They had to be focused on both glamour and nurturing. As TCA reported in 1938, "the stewardesses are a credit to Canadian womanhood both in efficiency and charm."[9]

"The passengers used to say, 'Oh, I love flying!'... But really they were just trying to be brave about it," Lucile remarked.[10] There was no radar, and no way to de-ice planes in winter — the airplanes flew low and the flights were rough. Many passengers were nervous. Everyone needed to put on an oxygen mask at high altitudes; the cabins were not pressurized. Pat once fainted when she took off her mask to persuade a passenger to put on his, but that only happened once when the flight was bumpy.

The planes were tiny — there were nine seats for passengers and one seat for the stewardess. Usually the passengers were men. The stewardesses handed out coffee, box lunches, lemonade, chewing gum, and razors, and they always carried a satchel stocked with Alka-Seltzer, cigarettes, and matches.

If the weather was rough the planes couldn't fly. The stewardesses had to arrange card games or parties at airports for the passengers while they waited for the weather to clear. Once Lucile had to ride by dog-sled, "bumping along on the frozen ice" to the local radio station at Fort Nelson in northern British Columbia during a weather delay. She could hear wolves howling in the distance but she thought, "This is great!"[11] During one flight in 1939, just to make the voyage "a little more pleasant," the male passengers from the plane were taken into a nearby field to pick fresh strawberries and the "obliging farmer supplied fresh cream."[12]

On one flight a stewardess had to find oxygen for a bowl of prize goldfish one passenger brought on board at Moncton. The fish began gasping and flopped over on their backs. Stewardess Mary Peacock Campbell rescued them by carrying the bowl to the nearest oxygen outlet, but that was later as more stewardesses were hired.[13]

Lucile and Pat first worked on a fifty-minute flight from Vancouver to Seattle. As the airline expanded, the flights became longer. They flew from Vancouver to Montreal, but with seven stops: Vancouver-Lethbridge-Regina-Winnipeg-Kapuskasing-North Bay-Toronto-Montreal. The trans-country flight was advertised as a "fast, daily, time-saving service ... [in]

the world's fastest commercial plane.... People thought it was wonderful to have breakfast in Vancouver and get to Montreal in time for dinner the next day," Pat remembers. "That was less than half the time of the fastest train journey."[14]

Lucile left the airline in 1941 to start a stewardess program at Yukon Southern Air Transport — later Canadian Pacific Air Lines — but resigned from that company at age thirty-two to marry Norman Dennison, an aeronautical engineer she met at Trans-Canada. They moved to Addis Ababa, Ethiopia for a short time with their two children, Margaret and John, where Lucile taught secretarial studies at a commercial school and Norman trained local airline staff in aero maintenance. After Norman's death she married the widower Lieutenant Colonel Jack Grant.

In her retirement Lucile kept busy gardening, bicycling, and enjoying her friends and cultural events. "I didn't think I was doing something daring by flying," Lucile said. "It wasn't until twenty-five years later that I realized it had been significant.... I always felt like a moth that had been released out of the closet and had her wings dusted off."[15]

CP Air also had stewardesses in its early years. In 1942 two nurses were hired — Jewel Butler and Melba Tamney — for the Whitehorse-Edmonton-Vancouver run. They looked after businessmen and, sometimes, pregnant women flying south for their delivery.[16]

The major airlines were not the only aviation companies to hire stewardesses. The historian Danielle Metcalfe-Chenail reports that in 1939 Jessie MacLean began as a stenographer for the northern airline Yukon Southern out of Whitehorse and soon became that airline's first and only stewardess.

If a woman wanted to fly and wasn't able to become a pilot, becoming a stewardess was seen as an exciting option.

10

Margaret Fane Rutledge and the Flying Seven

On a sunny day in June 1940, with a brisk southeast wind, four female pilots dropped more than 100,000 pamphlets over the city of Vancouver. As the papers fluttered down from the planes, sirens wailed to draw attention to them. The pamphlets, named "bomphlets" by the newspapers, had war messages such as "Give dimes or dollars to buy our men more planes."

This was part of an effort by some of the members of the "Flying Seven" to help win the Second World War. That day the women raised $100,000, which paid for eight new training planes. Imperial Oil had donated the oil for the planes the women flew, and the men at the airport helped get the planes ready.

Margaret Fane Rutledge (1914–2004), a bookkeeper and stenographer, had organized the Flying Seven in 1936, a small group of seven female pilots who, as women, were not even allowed to be spectators at most air shows. Margaret had met with Amelia Earhart in the United States to discuss how the few female pilots in British Columbia could work with other women and encourage more of them to become pilots.

The Flying Seven from left to right: Tosca Trasolini, Alma Gilbert, Jean Pike, Elizabeth (Betsy) Flaherty, Margaret Fane Rutledge, Rolie Moore Barrett Pierce, and Elianne Roberge, in Richmond, B.C., 1936.
Photo by Major James Skitt Matthews, 245-E-16, Vancouver City Archives.

The Flying Seven's first activity was a dawn-to-dusk "air patrol" publicity stunt. On October 15, at exactly 6:59 a.m., one of the women climbed into a small airplane cockpit, flew into the rain and fog over Vancouver, and stayed up for twenty-five minutes. The others followed in turn, flying for eleven hours in total with one plane in the air all the time, circling Vancouver. They wanted to make a point, one of them said, that contrary to public opinion, "a woman's place is in the air." Margaret was called a "daring young doll in her flying machine."[1]

The Flying Seven sponsored flying competitions and took part in air shows, mainly to persuade other women to become pilots. They met once a month wearing a grey uniform they designed with a crest and pin. Crowds of people were attracted to what were considered their daring stunts in the air.

But it was the Flying Seven Auxiliary School, which they organized in 1940, that may have made their greatest contribution to Canada. The seven women had applied to fly planes for the Royal Canadian Air Force during the Second World War, and the RCAF invited them to the force's headquarters in Ottawa thinking that they were men, but when they found out they were women they offered them jobs as cooks instead.[2] In response, they set up a school training fifty women to pack parachutes and work with fabric on airplanes, teaching them about flying. Many of these female "students" went on to work in factories making planes for the Second World War.

Margaret Rutledge had her first airplane ride when she was fourteen, deciding then and there that she wanted to fly. Her parents had flown in the first airplane to land in Edmonton when Margaret was six. Her father, who owned an automobile repair shop, even built his own glider.[3]

Margaret earned twenty-two dollars a week. Lessons were twelve dollars an hour, so she did the Edmonton Flying Club's bookkeeping and stretched fabric over airplanes' wooden frames in return for flying lessons. In 1933, at age nineteen, Margaret earned her private pilot's licence. Two years later she was the first woman in western Canada to receive her commercial licence. She had been taught by two famous pilots, the legendary First World War flying aces, Manitoba-born Wilfrid Reid "Wop" May and Maurice "Moss" Burbridge from England.

That same year Margaret also got her radio operator's licence, and is considered to be the first woman in the world to do so. In 1938 a small bush company airline hired her as their radio operator with an office at Zeballos, a small mining town on Vancouver Island. There were 1,500 miners and loggers in that town, and three unmarried women. Margaret was one of those women.[4] The story of her appointment was leaked to the press, and newspapers across the continent ran articles about her. The headline of an article by the *Chicago Daily Tribune* read, "Canadian woman pilot operating a radio station."[5]

In a navy blue uniform with gold braid and brass buttons, Margaret scheduled nine flights a day, looked after freight waybills, tied up and fuelled aircraft, and herded loggers and miners on and off the planes — sometimes because they were drunk. "Nobody thought anything about leaving … (gold) around loose," she said, "but you'd never leave a case of whiskey unattended."[6]

Margaret considered herself lucky. She often had an opportunity to fly as co-pilot when the airline was short of pilots. She also worked for Bridge River and Cariboo Airways as a dispatcher and occasional co-pilot. When her boss was "indisposed," usually meaning drunk, she flew the planes herself to isolated communities in British Columbia. Later Margaret became superintendent of reservations and traffic training for Canadian Pacific Air Lines, an excellent job, but affording no opportunity for her to fly.

The Margaret Fane Rutledge Family Foundation, created in her name, offers an award each year to a business administration student at Capilano University.

11

Queen of the Hurricanes: Elsie MacGill

Elizabeth "Elsie" Muriel Gregory MacGill Soulsby (1905–1980) was the first Canadian woman to get a degree in aeronautical engineering and the first woman in the world to design an airplane that flew, but sadly she was never able to become a pilot. When she was twenty-four she became ill with polio. After that she needed two canes to walk.

Elsie was brought up to believe that women could do anything. Her mother, Helen Gregory MacGill, was the first female judge in British Columbia. Both Elsie's grandmother and mother worked hard to get women the right to vote in elections in Canada as part of the women's suffragist movement. Elsie's mother owned a cottage near Vancouver, designated for women only; no men were allowed. There, Elsie learned to chop wood, clean fish, and repair broken windows and water tanks. She became known as "Miss Fix-It."[1] Her art teacher was Emily Carr, then a struggling West Coast artist. She also spent time with some of Canada's other famous suffragettes, such as Nellie McClung.

Elsie was the first woman in engineering classes at University of Toronto in 1923. "My presence in the … classes certainly turned a few heads," Elsie wrote. "The professor changed a lot of the words in his lecture when he discovered a woman was present."[2]

After she graduated with a degree in electrical engineering, Elsie worked for the Austin Automobile Company in Pontiac, where she began studying for her master's degree at the University of Michigan. It was there that she woke up one morning paralyzed from the waist down with acute infectious myelitis, more commonly known as "polio." A wedding was scuttled; she wrote her exams from a hospital bed. For the next three years she was confined to a wheel chair or her bed.[3]

The doctors told Elsie that she would never walk again, but she forced herself to learn to walk with two metal canes. She wrote magazine articles about planes and flying to help pay for her hospital bills and to fund doctoral studies at the Massachusetts Institute of Technology (MIT) in Cambridge.[4] Later she slipped on a rug and broke her one good leg. Again she forced herself to walk. Again she spent her time writing while she was recovering.

In 1934, hired by Fairchild Aircraft Limited in Longueil, Quebec, as an assistant aeronautical engineer specializing in stress analysis, Elsie helped design the first all-metal aircraft built in Canada.

Engineer Elsie MacGill.
Courtesy Canada Aviation and Space Museum 16035a.

Four years later she designed the Maple Leaf Trainer II airplane for training pilots. She couldn't test the plane herself, but she asked to be carried on board all the airplanes she was involved with for all the test flights — even the first flight, the most dangerous. It was said that Elsie flew from the rear seat of the Maple Leaf trainer.

When she was thirty-five, Elsie was put in charge of 4,500 workers, many of them young women, who built about 1,450 Hawker Hurricane planes at Canada Car and Foundry Co. (Can Car) for the British Royal Air Force to use in the Second World War. The Air Force needed lots of planes, and they needed them quickly. Elsie had to turn Can Car into an airplane assembly line almost overnight. From the thousands of blueprints sent from Great Britain, Elsie had to design the machines that would make the different parts for the airplanes. She also had to train all the workers at the factory. In 1942 a comic book was written about her life, calling her "Queen of the Hurricanes."

The next year Elsie married the works manager at Can Car, Bill Soulsby, a widower with two children, Annie and John Frank. However, everyone still called her Miss MacGill — even her husband. Both Bill and Elsie were fired from their jobs; workers weren't allowed to have an office romance. Instead they set up an office in Toronto, advising other companies on aeronautical engineering.[5]

When she turned fifty, Elsie wrote a biography of her mother, *My Mother the Judge: A Biography of Judge Helen Gregory MacGill*. Her mother had been heavily involved in "women's issues" and Elsie followed in her footsteps. She became president of the Canadian Federation of Business and Professional Women's Clubs and was appointed one of seven commissioners on the Royal Commission on the Status of Women in Canada. She advocated for paid maternity leave, daycare facilities, and liberal abortion laws.

Richard Bourgeois-Doyle, who researched and wrote about Elsie in 2008, describes her as "human, haughty at times and perhaps blind to the social context of her early engineering career.... For most of her life Elsie did not recognize gender bias and ... chose to ignore any hurdles in her career.... Her position was that ... one only had to 'knock harder.' But overall, I think her story should be something Canadians share and hold onto for inspiration," he wrote.[6]

Later in life, however, Elsie did change her way of thinking, absorbing much of her mother's approach. "Canada, too," she wrote, "has a caste system of a sort. Before our nation can marshall [*sic*] her full resources of ability and skill, Canadian traditional attitudes toward women must change, the present waste of woman power must cease, the reservoir of knowledge, skill and ability in the female half of our population must be brought into full usefulness."[7]

Elsie received several international awards for her work, as well as the Order of Canada in 1971. Four universities awarded her honorary doctorates. In 1946 she became the first woman technical advisor to the United Nations International Civil Aviation Organization, writing the International Air Worthiness regulations for the design and production of commercial aircraft.[8]

Dr. Lorna Marsden, former York University president, describes Elsie as one of the most important individual women of her time, "a woman who entirely changed the nature of her country, legally, economically, and certainly in terms of the quality of life."[9]

"I have received many engineering awards," Elsie said, "but I hope I will also be remembered as an advocate for the rights of women and children."[10]

12

Flying Blind: The ATA

Marion Alice Powell Orr (c.1918–1995) was lost, flying over England in clouds and rain. Her fuel was almost gone. She saw a river below and decided to ditch her plane in the water to avoid hitting a house. But as she came down, she spotted the orange lights of an airport. Marion slipped safely to the runway below. On another flight, flying north along the coast, she couldn't see the ground. Suddenly she was so low over the ocean that she could feel the spray and taste the salt. Somehow she made it back to her base.

Marion was one of five Canadian women flying for the Air Transport Auxiliary (ATA) — a British organization that hired male pilots who were too old or for some other reason not allowed to fly for the Royal Air Force (RAF) — during the Second World War in Great Britain. The idea was to free able men for combat duties. Soon there was so much work that female pilots were hired too, working on an almost equal footing with the men, doing what had earlier been thought of as strictly men's work.

Marion and Violet "Vi" Milstead Warren (1919–2014) had been flying instructors in Canada, but gas was rationed during the war from November 1942 on, as much of the gas was being used to fly military planes. Flying schools weren't allowed to teach non-military pilots, so Marion and Vi were out of work. Fortunately they were hired by the ATA. In April 1943 they crossed the Atlantic Ocean first class, with their steamer trunks on a troop ship in convoy.[1]

For the next two years, as part of the ATA, their job was to fly military airplanes from factories to Air Force bases or other safe hiding places, readying them for the RAF pilots. The Air Force wanted the airplanes out of the factories as soon as possible in case the enemy bombed the factories. Sometimes the airplanes were even stored in orchards. The ATA pilots also flew damaged aircraft to the factories for repair, or in some cases to the junkyard. As Vera Strodl Dowling said, "There were planes that came in shot to pieces and we had to pick them up and take them to the dump as it were, to be repaired or used as parts."[2]

It was hard work, and dangerous. One hundred and seventy ATA pilots lost their lives during the war. Some of the planes they had to fly were so damaged that they were only serviceable for one landing. As Vi put it, "You ran into difficulties.… You got out of them and there was nothing much to tell. Or you didn't get out of them and someone else did the telling."[3]

Each female pilot made up to eight flights a day on four or five different aircraft. They worked two weeks straight, then with four days off, doing the same work as the men but with twenty percent less salary.[4] The women had little training. They were given a pocket "Blue Book" of small index cards for each type of plane describing how to take off, fly straight, and land. Sometimes they were taxiing down the runway for takeoff as they read the instructions for that particular plane. "There were a lot of knobs and dials you just ignored," Vi confessed.[5]

"All planes are fundamentally the same," Vi wrote. She related a conversation she had with a male pilot as she was climbing into a type of plane she'd never flown before:

> Male Pilot: "Have you got much time on this type?"
> Vi: "No. I've never flown one before."
> Male Pilot: "Never flown one before?!" (*in an outraged voice*)
> "Then what makes you think you can handle it?"
> Vi: "No problem. It's all in my book."

Male Pilot: "In your book! Good God, girl, you can't fly this from a book!"

Vi: "I can — from *my* book."[6]

The pilots flew by dead reckoning. Not allowed to make radio contact with anyone in case the enemy was listening, they had to fly by reading maps, watching a compass, and looking out the cockpit window. It was tough when you were only five feet tall like Vi. She often sat on her black leather overnight bag in order to see out.

The roads and railway lines were hard to figure out. Marion thought the roads looked like spaghetti — "roads all over" instead of running north and south.[7] There was often smog, fog, rain, and smoke from factories, making it difficult to see the ground. It was said that ATA pilots flew when the weather was so bad that "even the birds were walking."[8] There were "barrage" balloons — huge balloons anchored to the ground to keep enemy planes away. The pilots had to avoid them as some of the balloons would explode if you hit them. There were dummy airports to fool the enemy; they had to be able to recognize those.[9]

While the ATA motto was *Aetheris Avidi*, "Eager for the Air," ATA pilots said the organization's initials really stood for "Anything To Anywhere." During the war the 1,318 ATA pilots altogether made 309,011 flights. They flew ninety-nine different kinds of planes. Marion Orr's favourite was the Spitfire, which she considered light, graceful, and easy to handle — "the most beautiful plane ever built."[10]

One hundred and sixty-six of the pilots were women. There was, of course, opposition to this venture. C.G. Grey, the editor of the magazine *The Airplane*, wrote: "The trouble is that so many of them (female pilots) insist on wanting to do jobs they are quite incapable of doing. The menace is the woman who thinks that she ought to be flying a high-speed bomber when she really has not the intelligence to scrub the floor of a hospital properly, or who wants to nose round as an Air Raid Warden and yet can't cook her husband's supper."[11]

One of the Canadian women, Vancouverite Helen Harrison Bristol (1909–1995), claimed "it was the most exciting and interesting time" of her career. Helen had taken her flying lessons in England when her family moved there. After her experience as an ATA pilot, she returned to Canada only to find that she couldn't get a job in aviation. Trans-Canada Airlines, Pacific Airlines, and Canadian Pacific Air Lines told Helen that she was too weak to handle the controls. She became a taxi driver at Dorval airport in Montreal instead. "I had 2,600 hours flying civil and military aircraft in three countries, and they gave the job to a man who had only 150 hours," she said. "I was furious. I just couldn't believe it."[12]

Helen later instructed pilots in float planes in British Columbia; her reputation as a competent instructor inspired one of her students to compose a poem, which ended:

> There are some good teachers,
> Professors and preachers,
> Who'll give a thing all that they've got,
> And in spite of her yellin'
> When you're taught by Helen,
> You'll sure as Hell know you've been taught![13]

NEW OPPORTUNITIES

13

Into the Woods: Bush Pilots

Although an advertisement in a 1955 issue of *Maclean's* declared, "Attractive careers and important duties for Women in the RCAF ... the same pay ... the same rank and the same opportunities for advancement as airmen...,"[1] the reality was that there were very few jobs for female pilots after the war, although small airlines that flew up north and into other remote areas sometimes hired women as bush pilots.

Bush planes and their pilots were important. It was the only way some people up north could go to doctors quickly, or get their groceries and other supplies. It was the only way that trappers and miners and geologists could get in and out of out-of-the-way places in a timely fashion. But it was dangerous work. The area the pilots flew into was inhospitable with rough terrain. Often there were no prepared landing strips or runways. In winter many landmarks were covered in snow; everything was white. Many times pilots flew rescue missions in danger of needing rescue themselves.

Male bush pilots often resented female pilots, believing that the women weren't strong enough to lift the cargo and the men would have to do double work. Men also felt that the women were taking jobs that should have been given to them; many men still felt that flying planes was a man's job. Women in this arena were very much second-class citizens.

One of the male pilots once said to Lola Reid Allin, "The problem is that we don't know how to treat you … whether we should open the doors for you … or if we should just walk through as if we were with another male pilot…"

Sometimes the men played practical jokes on the women — partly to scare them away; often just for fun.

Lorraine Cooper (1931–2006)

Lorraine R.M. Cooper, a tall attractive brunette, was eager to do a good job for Chinook Flying Service. One day after a fierce snowstorm she was told to fly into the bush to pick up Mr. Stiff at an isolated farmhouse. They said he needed special treatment.

Lorraine arrived at the farmhouse and looked for Mr. Stiff. "Where is he?" she asked. "Over there," a man replied, pointing to a canvas-wrapped bundle in the shape of a body, outside in the freezing cold. Only then did she remember that dead bodies were often called "stiffs." "I wanted to run," Lorraine said, but she picked up the bundle and put it into her small plane. She had trouble getting the body in because it was so hard and too long. She had to rest the feet-end on her shoulders.

When she returned to her base, all the male pilots were watching. "Hey guys," she said, trying to smile, "that was a good joke." After that, they didn't play many practical jokes on her.[2]

Before working for this company Lorraine had trouble finding a job as a pilot. She used her name with an initial, "L. Cooper," to apply; thinking it was a man, an airline would give her an interview, but when she got to the interview and they found that she was a woman, they refused to hire her. At one point she washed airplanes and swept floors in return for being allowed to fly.

Lorraine sketched and wrote *haiku* about nature. Her *haiku* of Canada, "Where/Wind Waves/Wander," were often about the sky:

> Look, geese are writing
> A strange alphabet on the
> Blue parchment sky.

Broken sky mirror
Lo! Heaven's fallen pieces
Blue forget-me-nots.[3]

Lorraine changed her career, going to the United States to work in marketing and research. She came back to Canada in 1995, where she opened her own consulting firm in Victoria, B.C., and then went back to Calgary. She became an active volunteer for several organizations.[4]

Violet "Vi" Milstead Warren (1919–2014)

A slim, petite woman, Violet "Vi" Milstead Warren was officially the first and best-known Canadian woman bush pilot.[5] She worked with Leavens Brothers as an instructor, and then with Nickel Belt Airways in Sudbury, just after she came home from flying with the Air Transport Auxiliary in Great Britain (see chapter 12). Her husband, Arnold Warren, became the chief flying instructor with Nickel Belt; Vi's main job was running the flying school, but she managed to get some flying time too. This was often the way women got their foot in the door — being part of a husband and wife team.[6]

Vi flew mine surveyors, trappers, lumberjacks, hunters, fishermen, and men who were fighting forest fires. As one male pilot said, "There is hardly a trapper, bushman or prospector ... that does not recognize this small parka-clad figure at the controls" of a plane.[7]

The well-known Canadian journalist and social activist, June Callwood, described Vi as a "Bush Angel." June took flying lessons from Vi in 1946 at Barker Field on the outskirts of Toronto, and June was so impressed by Vi that she wrote an article about her for a women's magazine. June loved the "feeling of peace and quiet as she flew up into the sky, away from life's troubles." But when she almost hit some power lines at dusk, June decided that, as a new mom, she should stop flying.[8]

Vi and Arnold worked in Windsor, Ontario, and Arnold worked for the International Civil Aviation Organization in Djakarta, Indonesia. There Vi was given a piece of paper saying that she could fly in Indonesia, but they wouldn't let her teach flying. By the end of 1954 they were both back in Canada, Vi as a librarian and Arnold working at a community college.[9]

Vi had her share of difficult co-pilots and passengers. One male pilot tried to kiss her as they were about to take off. She spun him around and kicked him out the door. Fortunately they were still on the runway. A male passenger said that he had never been kissed in the air. She said "I have and it's no different from being kissed on the ground." One writer claims that a miner was making sexist comments to Vi for hours, and when she landed the plane she kicked him out the cargo door into the lake. Perhaps these are all versions of the same story.

Vi learned to fly in Toronto in 1939 and received her pilot's licence that same year. She had decided to fly when she was in high school, when a little plane buzzed their playing field during a football game. Earlier she had wanted to be a surgeon, but her mother needed Vi's help to open a wool shop and took Vi out of school; Vi saved all the money she earned to pay for flying lessons. Later she took charge of the wool shop herself.

She flew at Barker Field, north of Toronto, at 7:30 a.m. for a half-hour lesson in order to be at work at 9:00. After six months of lessons she had both a pilot's private licence and a commercial one. Her instructor made a film of her flying, "Let's Learn to Fly."[10] After the war Vi flew again at Barker Field, instructing pilots. One of her more exciting tasks was flying the 1946 Miss Canada to Washington to invite the president of the United States to Canada for Canada's first international air show. Vi also met her husband, Arnold, at that airport.

When Vi flew she said she "felt alive in a way I had never imagined" … [I] "pity the earthbound."

Vi also became active with volunteer organizations including Meals on Wheels and Rotary initiatives. She was awarded the Order of Canada in 2004, an Amelia Earhart Medal, a Paul Harris Medal, the Rusty Blakey Memorial Award, a Diamond Jubilee Medal, and was inducted into Canada's Aviation Hall of Fame. A special postage stamp was issued in 2009, honouring her career.[11]

Lorna Bray Nichols de Blicquy (1931–2009)

When she was fourteen, after a cousin had taken her for a flight over Ottawa, Lorna de Blicquy announced to her parents that she wanted to learn to fly. Obsessed with flight, she joined the Ottawa Parachute Club, and at sixteen became the youngest Canadian to make a parachute jump.[12] She wrote an essay in high school comparing flying to "a symphony in shining silver," although her English teacher didn't think much of the simile.[13]

Lorna's brother was an RCAF pilot and her parents weren't very pleased that she wanted to be a pilot, too. Nevertheless they agreed to drive her to flying lessons as long as she "had a 75 per cent average in school, with no marks below sixty." She had to take an early bus home, as her mother believed that "nice girls don't hang around airports."[14] At one point Lorna was writing to her brother asking him to tell their parents how wonderful it was to fly, while at the same time her father was also writing to her brother asking him to dissuade Lorna from flying.

Lorna worked hard to scrape enough money together for her lessons, babysitting, working as a cashier at a movie theatre, and doing all kinds of odd jobs at the airport. At fourteen she took her first solo flight in a light yellow plane. Lorna got her pilot's licence in 1948, went to Carleton University in Ottawa, and during one year in 1953 she received her B.A., her teaching certificate, her commercial pilot's licence, and married Tony Nichols.

Although a skilled pilot, Lorna had a difficult time finding a job, so she taught high school and served as a flight instructor in her spare time. When she applied to be a bush pilot it caused a great deal of laughter in the northern bush camps — that a woman had applied. But she got her break one day when a regular pilot didn't show up and she was allowed to sub for him.

Sometimes men wouldn't fly with the female bush pilots. On one such occasion a male school inspector didn't want to fly with Lorna because she was a woman. He waited for three days for a male pilot, but none showed up. Finally he had to fly with Lorna. She took extra care when she landed, staying far away from the rocks at the dock, wanting to make sure the school inspector wasn't worried that she would hit them. Regardless of her efforts to impress, he criticized her for landing too far away from the dock.[15]

Lorna had been turned down for so many jobs when less qualified male pilots were hired, that she often complained on radio and in newspaper and magazine articles about this discrimination. When she had 6,000 flying hours, she applied to Air Transat, but was not even granted an interview on the basis that she was so short, they would have had to design a special uniform for her. Two of her less qualified male students were hired instead. She charged the company with discrimination.

Some of Lorna's articles received national attention. A guest editorial in 1974 in *Canadian Flight*, protesting this discrimination against female pilots by the Crown Corporation Air Transat, attracted other media attention and was deemed partly responsible for major airlines later opening their cockpits to women.[16] Lorna gave her male colleagues praise, though. "Men will usually help you — unload your plane or refuel it — but you need to know how to look after yourself," she explained.[17]

Over thirty years Lorna flew in the Arctic as far north as Resolute Bay — the first woman to fly there commercially — the Northwest Territories, eastern Canada, New Zealand, the United States, and Ethiopia. At one point she lived up north in a tent for more than two years. While she was flying in New Zealand, a propeller came off her plane, but she landed safely.[18] The plane had been defective.

Retired pilot Lorna Bray Nichols de Blicquy at her Pioneer Hall of Fame induction in 1996, with Karen McArdle from the Women in Aviation International Board standing behind. *Women in Aviation International.*

Lorna was an instructor, charter pilot, helicopter pilot, bush pilot, flight test examiner, and Canada's first female civilian aviation flight inspector. She flew on floats and skis, as a helicopter pilot and as a glider pilot. She taught English in high school and people how to fly. She had trained to be a helicopter pilot in the late 1960s, when she was in her forties, explaining that, "you can't stand still in this business."[19]

Lorna won numerous awards and honours, including the Governor General's Award and the Trans-Canada McKee Trophy in 1993, and was inducted into the Pioneer Hall of Fame in 1996. She served on a committee for pilots' medical standards on pregnancy. "She absolutely adored flying," said her daughter Elaine, "so she was always amazed that people wanted to reward her for stuff that she thought was fun."[20]

"I have had the best flying life possible for one who never wanted to do anything else. I am lucky, thankful and pleased to have spent most of my life doing what I loved," she wrote. "I don't know what I'd do if I didn't fly," she said. "I'm really a boring person," although she read, swam, and played tennis and bridge. Lorna stopped flying in 1999.[21]

On hearing of her death in 2009, one blogger wrote: "Good-by Lorna. Many women will thank you for the inspiration you provided to be a pilot. You sure were one tough lady, but in your time, you had to be. Strangely it is still necessary for a female pilot to be three times better than a male in order to get respect."[22]

The Canadian Ninety-Nines issued a commemorative postage stamp honouring Lorna in 2011.[23]

Elizabeth Wieben

Elizabeth Wieben grew up in Thunder Bay. Orville Wieben, her bush pilot father, treated Elizabeth and her sister the same as he treated his three sons. They all were encouraged and supported in their journey to learn to fly. The girls were expected to do the same hard work as the boys, even to brave the bitter cold of the north.[24] The whole family ran an air charter business in Thunder Bay, Superior Airways Ltd.

As a special treat, when Elizabeth was little, her father used to take her with him when he was flying in and out of camps, but eventually she got too old to bunk in camps where the men stayed; there were no facilities for women.

By the time Elizabeth finished high school she already had her commercial pilot's licence and her instructor's rating — she was qualified to teach flying. Later she would also earn an Australian commercial licence, an American commercial licence and instrument rating, and a Canadian airline transport licence.

After she graduated from university she was hired as manager of the Lakehead Flying School in Thunder Bay. There she met and married Robin Webster and they went off to Australia for four years for Robin's work as a geologist. She had two boys there, Colin and Kenneth, and when they came back to Sudbury their daughter Roberta was born. They had another son, Donald, in Washington State. Elizabeth kept busy during these itinerant years. She spent her time writing exams and collecting pilot's licences from different countries, as well as "having babies, and flying."[25] When they finally returned to Canada, to the north shore of Lake Superior, she and her husband Robin operated their own air charter and outfitting business, Wieben Air Ltd.

Bush pilot Elizabeth Wieben in the family DeHavilland Beaver DHC-2 C-FODN in 1986. *Robin Webster.*

As a bush pilot Elizabeth flew in fog, rain, and snow, often in below-freezing weather. Many times she flew alone, without radio contact. She had to deal with unruly passengers, and sometimes she had slippery loads of fish. One time when she was hauling fish on her own, someone else packed them on the plane in the small section just behind her seat. When she had to stop the plane quickly, the load of slimy, scaly fish landed down her neck and in her lap.[26]

Only five-feet, two-inches tall, Elizabeth once had to lift her float plane all by herself off a rock. Often she'd take her kids flying with her because she needed their help, or because there was no one else to look after them.

There were emergencies to cope with — most as the result of chain saw or axe injuries. One man she picked up had a chain saw injury to his neck. In another instance a man wanted to get out of the bush earlier than his scheduled pick-up date, and the only way he could think of to get her attention was to stand naked on a rock beside a lake. It worked.

Elizabeth got a job as a co-pilot in 1976, with the condition that she got on the plane before the passengers, stayed on until they got off, and didn't use the intercom on the plane. The airline didn't want the passengers to know a woman was one of the pilots.

Elizabeth taught her husband to fly; he was one of her students. Later they opened their own aviation company where she worked full-time while raising four children.

She taught flying at Confederation College in Thunder Bay, Ontario until 2005 as professor, aviation program coordinator, and director of flight operations. She not only taught students how to fly, but also instructed in subjects as varied as meteorology and aviation law. When the geologists from Falconbridge Mining Company heard that she'd be teaching, they found out when her off-days were and decided to fly only on those days when she was the pilot.

When Elizabeth retired she was honoured with an outstanding service award and a designation of professor emeritus. She is a founding board member of the Northwestern Ontario Aviation Heritage Centre and received the prestigious Northern Lights Award in 2013.[27]

Elizabeth has worked as an accomplished bush pilot, flight instructor, air charter operator, flight test examiner, and civil aviation tribunal officer. "A few of the male pilots were critical or mean-spirited," she admits, "but there were many others who provided real support."[28]

Judy Adamson

Another bush pilot, Judy Adamson, points out that fish hauls were always smelly. Even though the pilots spread plastic over the floor of the plane to contain the fish, and scrubbed with detergent afterwards, it was impossible to get rid of the fishy smell.

Judy worked for Parsons Airways Northern in the late 1970s, and as her boss wrote, "There was nothing glamorous about her job. She hauled smelly fish and flew people and equipment into mining and fish camps. She always went out of her way to do her share."[29] It was while she was flying in northern Manitoba that she met the geologist she would later marry.

But Judy didn't always haul fish. In 1982 she became the first woman to fly a Canadian CL-215 water bomber. When she flew to drop the water, she had to fly as low as one hundred feet above the trees. The work was dangerous, as flames from the fire could reach that high in the air, close to the plane.[30]

"Working the fire is physically and mentally exhausting," Judy says, "for it means continuous landings and take-offs to scoop up and drop water." The bombers pick up water at almost a ton a second, and in about ten to twelve seconds the tanks are full. The planes land on the water and maintain their speed without stopping, fill the tanks, then quickly fly up to drop the water. Judy was one of the few women strong enough to cope with the task; she has always been an athlete and skis expertly.[31]

Judy flew water bombers for five years. She spent twelve and a half years co-chairing committees and in other positions for Transport Canada. Then in 1999 she was hired by Air Canada as a pilot and flew until she retired at the end of 2011, having flown throughout North and South America, Europe, and Asia. "It was a wonderful career," Judy reminisces. "My best memory was flying with Judy Cameron in Dreams Take Flight, a program in which airlines provide the planes and pilots volunteer their time to take medically, emotionally, physically, and socially challenged children to Disney World in Orlando for a fantastic fantasy-filled day."[32]

Judy Adamson flew water bombers to help put out forest fires before she became an Air Canada pilot.
Courtesy Judy Adamson

Berna Studer McCann

Berna McCann grew up in a log cabin in northern Saskatchewan, one of seven children. Her father was a farmer, prospector, trapper, and fisherman, and they could get to their home only by dogsled in the winter and canoe or plane in the summer. All of the children were fascinated by the airplanes that flew over their house. If a plane landed, they would "pester the pilot for a ride or to sit in" it.[33] They made model planes and lived under the assumption that one day they would become pilots themselves.

Berna worked in a bank in Prince Albert, earning her pilot's licence there when she was twenty-one. She wore a pink flying suit and she and her sister Joan flew all over the north, sometimes in Berna's pink Cessna airplane. They even went to the Northwest Territories to look for gold, having learned how to recognize different rocks. They both loved camping and fishing.

Bush pilots were expected to work long days, often without a break. Berna was used to hardship, but she found the hours as a bush pilot in the 1960s long. She worked fourteen to eighteen hours a day, seven days a week, hauling groceries, supplies, and people. "A person gets run down from lack of food and sleep," she said.[34]

Berna had been hired as a pilot because of "her ability to live in the bush and look after passengers in the event of a forced landing." But the male pilots didn't all agree with the decision to hire her "in what was looked on as a man's game."[35]

Flying bush planes "was a hard way to make a living," Berna said. In 1970 she opened Tamarack Lodge, her own tourist camp in the north. It was so isolated that she had to fly out to buy groceries, as she didn't own a car. Eventually she bought one and drove instead.[36]

Judy Evans Cameron (born 1954)

Air Canada's first female pilot in 1978, Judy Cameron also began her flying career as a bush co-pilot. She flew in sub-zero weather but spent a lot of her time loading and unloading cargo. She wore coveralls and steel-toed boots. "I put muscles on muscles," she laughs, although she claims that one of the most physically demanding jobs she ever had was flying a DC-3 out of Inuvik.[37]

When Judy was young her mother taught her that she could be anything she wanted to be. The first time she flew in a two-passenger plane the pilot asked her if she wanted to see a pencil float. As the pilot performed an aerobatic manoeuvre, he floated the pencil from the front of the aircraft to the back; Judy decided it was so much fun that she wanted to become a pilot.

After she went to university, she bought a motorcycle; it was cheaper than a car, and her mother would ride around with her on it. Later she sold the motorcycle to pay for flying lessons. Judy went to Selkirk College in British Columbia where she studied to be an airline pilot, the only woman in her class. "I had discovered that women were not universally accepted in all fields and certainly flying was considered to be for men only."[38] During the summer she worked on a road crew, digging ditches and planting trees.

Judy was hired by AirWest, which paid her $450.00 a month as a part-time co-pilot and reservations agent, while the full-time pilots made only $400. It is not surprising that this caused some resentment.

She found another company that was brave enough to hire a female pilot, but their planes were old — the doors would fly open in the air and the landing gear would sometimes refuse to come down. Perhaps fortunately, the company went bankrupt.[39]

Judy finally got a job with Gateway Aviation, based in Inuvik. The chief pilot said that although "he didn't want to hire a woman, he would because he knew that I wouldn't leave him for the [major] airlines." But, of course, that was her dream — to fly for a major airline. She was on call twenty-four hours a day and sometimes had to make as many as six trips a day.[40]

As a co-pilot in a DC-3 Judy had to do all the fuelling. "I got really good at leaping onto the wing.... It seemed to take forever to fuel it … when you were up there with the mosquitoes." But Judy found the 45-gallon fuel drums heavy, and the captain wished he had a stronger male co-pilot.[41]

Judy Evans Cameron, first Air Canada pilot, now retired, screamed the first time she flew in a two-seater plane when she was nineteen. Later, when she was a pilot and in uniform, she was often mistaken for a stewardess; passengers asked her to carry their luggage or hold their baby while they went to the washroom.
Rick Madonik/GetStock.com.

When Judy became Air Canada's first female pilot, she endured a few negative comments. One passenger said, "A woman pilot. Well at least it keeps them off the roads."[42] Judy was selected by Air Canada out of 1,800 applicants. During the next two years after she was hired, Air Canada received 1,728 applications, twenty from women, of whom five were hired.[43]

In her first year with Air Canada Judy earned $1,100 a month — on the same pay scale as male pilots. Her uniform was virtually the same as that of a male pilot except that she wore a navy ascot instead of a tie and a "jaunty navy felt hat with a tiny brim."[44] Newspapers, radio, and TV kept asking her for interviews. Some of the senior captains suggested that their male pilot friends would not accept female pilots, but of course they would themselves, each pointing their finger at the others! In the end, Judy's good humour and pleasant personality finally won them over.

Judy was the first female pilot with Air Canada, and the first female captain of a Boeing 777. She flew thirty-seven years with the same airline, a record for women in Canada, and 21,000 hours with that airline, another record for women. She flew 23,000 hours in total.[45] Judy has two children, Carolyn and Kristy.

Lola Reid Allin (born 1955)

Women bush pilots proved to be capable and conscientious, but even as late as the 1980s some people found it hard to accept them. Discrimination could be found almost everywhere.

Lola Reid was about to sit down at the controls of a twenty-passenger-seat plane when a male passenger asked her what food she'd be serving on the flight. She explained that she was the co-pilot, not a flight attendant. She pointed to the stripes on her uniform indicating her rank. "Oh," he said, "I thought you had just borrowed the jacket."[46] Another passenger wrote a letter to the airline complaining that Lola had spent all her time sitting with the pilot, even holding his hand on takeoff, ignoring the passengers. The airline wrote back that they didn't have a flight attendant on that flight; Lola was the co-pilot.

Lola prepared for her flight tests while also working in the computer division of Canadian Imperial Bank of Commerce. Later she earned a BA

Bush pilot Lola Reid Allin, 1986.
Robert Wong, Our Town Archives.

in psychology while flying up north for Bearskin Air; she studied during layovers. She flew for ten years, both small land planes and floatplanes, acquiring 6,000 hours of flight time. She taught aviation ground school and flight simulator programs at Flight Safety International/De Havilland. During her career as a commercial pilot, her real passion was flying solo, performing acrobatic manoeuvres such as stalls, loops, and rolls.[47]

Interested in art and travel since she was six, Lola became an artist, writer, award-winning photographer, guide, and scuba-dive master, and she also enjoys weaving. Now she travels around the world to countries, such as Morocco, Guatemala, Mexico, and Kenya, taking photographs, painting with water colours, giving lectures, guiding, volunteering her time as a teacher and translator, and writing travel articles for such prestigious magazines and newspapers as *The Smithsonian, National Geographic, The Globe and Mail*, and *National Post*. Lola has always had a keen interest in anthropology, a subject she studied at university.[48]

While she loved flying, she says she doesn't miss standing out on an airplane wing, twenty feet above the ground in a blizzard, clearing ice in sub-zero temperatures. But she has many fond memories. "Sometimes when you see a sunset turning all the tops of the clouds a bright gold colour, it can be a spiritual experience," she says.[49]

Mary Ellen Pauli

Not all bush pilots fly small planes — there are those who fly helicopters. Mary Ellen Pauli is one such pilot.

Mary Ellen grew up in Matagami in a trailer in the north of Quebec, dreaming of becoming a pilot like Amelia Earhart and a bush pilot like her father. "There were not that many women flying, especially in the field I wanted to go into," she said. She heard discouraging words at every step along the way. One man said to her, "If women were meant to fly, the sky would be pink." Even her father told her that she'd never get a job.[50]

She had a difficult time getting trained to fly. "My mother had always said, 'Where there is a will, there is a way' and I think she was referring to getting housework done, but I took my own spin on it.… I just had that goal and moved forward," Mary Ellen said.[51]

When she was old enough, Mary Ellen got a job in an office. She saved $11,000 from her salary to pay for her pilot's training. Then she sent a cheque to a helicopter company in New Brunswick to reserve a place in their training program. When she arrived for her training, however, the manager said he wasn't expecting her. They were only training and only had accommodation for men. "Well, you cashed my cheque," Mary Ellen said to the manager. "I'll move in and if there are any complaints, I'll move out."[52]

There were no complaints and Mary Ellen got her licence, but when she applied to several companies for a job, they all said, "Our work is in the bush and women aren't allowed in these camps, so we wouldn't know what to do with you."[53]

Finally in 1980 she got the job she wanted with Trans Quebec Helicopters. The boss there didn't want her because she was a woman, but there was no one else and she quickly became well known and well accepted. She had to live in tents, trailers, and barracks where she was the only woman.

Now Mary Ellen flies for the Ministry of Natural Resources. The work is quite varied, "which is what I like about it," says Mary Ellen. "In the summer we do a lot of forest firefighting. Then come winter we do a lot of moose surveys [to find out how many there are]. In September that's when we're doing our polar bear projects [to track them and see where they go] ... plus we do timber management surveys."[54]

She's had her share of frightening moments, such as the time polar bears were lunging at her helicopter, and another time she had to yell at one bear to frighten it away. She was awarded a Governor General's Certificate of Commendation in 2004 when she rescued two adults and three children who were stranded on a small low-lying island at the mouth of the Sutton River in Hudson Bay. She flew her helicopter into strong winds, sleet, and rain to pick them up. She was honoured in the Ontario Parliament for this act of bravery.[55]

Mary Ellen has recently been featured in a special exhibition on "Women and Aviation" at the Bushplane Heritage Centre in Sault Ste Marie, and among her many accomplishments, in 2012 won one of the four annual Elsie MacGill Northern Lights awards for Canadian women in aerospace and aviation.[56]

Helicopter bush pilot Mary Ellen Pauli.
Copyright 2014 Katelyn Malo.

Mary Ellen has flown for mining exploration camps, power line construction, diamond drilling, environmental studies, and transportation areas, but she has other talents besides flying. Mary Ellen, her husband, and her two children, Cameron and Audrey, are all accomplished musicians and first-class skiers. Mary Ellen and the kids play the violin and the piano; John plays the bagpipes. Recently she took part in a European tour with the well-known Kitchener-Waterloo Symphony Orchestra. She's also an active volunteer in her community.

As her mother said, Mary Ellen had the will to find a way to be what she wanted to be.

14
Around the World: Daphne Schiff and Friends

Canadian pilots Daphne Line Schiff (born 1924), Adele Englander Fogle (born 1934) and an American friend, Margaret Riggenberg, were flying in an around-the-world race over Iran in the *Spirit of '76*, a little white twin-engine Cessna 340, when suddenly two large Iranian military jets appeared beside them — one on each side. Were they forcing the women to land?[1] Panicked, all they could think of was what they could fashion into a head covering for when they reached the ground, realizing that in Iran most women covered their hair. Quickly the three female pilots searched the little luggage they had carried onboard for something suitable; they could find only their underwear. Fortunately the jets left them alone after ten minutes. Relieved, they continued on their 1994 Round the World Air Race.[2]

It was an exciting and somewhat dangerous trip. Their radio stopped working when they were over the Atlantic Ocean; a thunderstorm put them off-course; they narrowly missed a typhoon; and their plane was freezing cold. They had removed three of the plane's six seats in order to carry a forty-gallon fuel tank, so they couldn't put the heater on in case the tank exploded. They had to sit in cramped quarters for up to twelve hours at a time.

But they found magic, too. Over Alaska they saw a circular rainbow with the shadow of their small plane in the centre of the circle. Twenty-five days after they began, they landed in Montreal, having flown over countries as diverse as Morocco, Turkey, Dubai, India, Vietnam, Japan, Russia, and the United States.

Daphne and Adele flew together in other races, "to keep up their skills." In one air race of the Americas, their engines stopped over the ocean about one hundred miles from Chile. They had dropped very close to the sea when they discovered they had water in the fuel tanks. Because they were skilled pilots, they managed to get the plane back up. "You have a checklist," explains Daphne, "and you know what you have to do — you go through it item by item until you find out what's wrong. And that's what we did … in the meantime, the ocean's getting closer," she smiles.[3]

When Daphne was young she wanted to study chemistry, physics, and math at university; the school said "Sorry, that's for boys." She took the courses anyway, and after earning her degree went to work at Chalk River, purifying plutonium. She earned a master's degree in chemistry in 1947.[4]

She didn't learn to fly until after she was married. Her husband, Harold, a science professor, had his pilot's licence; he felt it was safer if they had two pilots onboard when he flew their two kids to a camp in Quebec.[5] Daphne was terrified during her first lesson. She clutched the instructor's arm the whole time. "She'll never be a pilot," the instructor told her husband.

Pilot Daphne Schiff.
Courtesy Adele Fogle.

"That's when I decided to do it," Daphne relates. "My parents, both teachers, brought me up to believe girls could do anything."[6]

Soon she loved being in the air. "Flying gives you a chance to look at the world from" a different "point of view," she says. "There's a certain independence to flying, but it's beauty more than anything. I love weather, and being up in weather is exciting."

Daphne taught science at York University in Toronto, and was on the faculty for thirty-eight years. Eventually she taught the science of weather and flight, many years landing a Cessna plane on the university campus to demonstrate flying to her students. "I thought it would be great for the students to have a real aircraft to touch and examine. And, let's face it," Daphne says, "you need a gimmick to teach science," to keep the students interested. When she was teaching there every student needed to take at least one science course.[7]

Daphne also flew delivery planes for Purolator courier. She wanted to work for a major airline, but women weren't being hired. "You still hear stories nowadays about passengers who get on board an airplane and they're shocked to see a woman pilot," Daphne says.

Pilot Adele Fogle.
Al Gilbert C.M.

* * *

Adele Fogle grew up in a tiny apartment in Toronto, the youngest daughter of European immigrants. Her mother was Russian, her father Polish. "My sister and I slept in the dining room. My aunt paid half the rent so she got" a bedroom, Adele says. "My mother saved every penny she could and sent us to summer camp. I saw what those kids had and I wanted it, too."[8]

When Adele was thirty-seven years old, she went with a neighbour to his flying lesson, and became convinced that she wanted to fly too. Her two daughters were in high school and she was looking for something to occupy her time. She'd tried horse breeding and computer programming, but she wasn't fulfilled by either.[9]

When she began her flying lessons, she was the only woman in a class of men. "I was being taught by kids much younger than me and I had no idea about the basic lingo of aviation or the basic background knowledge that you require to do it … I had to push myself to keep booking flights."[10]

The first time she flew solo she crashed — "directly in front of the air-field cafeteria, which was filled with lunchtime visitors.… I wasn't at all frightened. I just figured, next time I'll do it better." She never mentioned the accident to her family or friends she was so embarrassed. "All of my friends thought I was crazy to be flying anyway," she admits. "But it's not that hard to do. Anyone can become a pilot. It will open up their whole world."[11]

Soon after that accident Adele broke her leg skiing. It was two years before she could fly again. "But I love to fly." It's such a "feeling of freedom and control," she says. "I won't give up. I persevere and persevere …"[12]

After Adele got her licence, the only pilot's job she could find was as a courier. She bought an old Aztec plane to use for the job. "I'd leave Toronto at midnight every night for Montreal to deliver the mail. Then I'd wait until 4 a.m. to return with the newspapers."[13]

She also helped organize a project sponsored by female Canadian pilots — taking underprivileged children on flights over Toronto's downtown when the Christmas lights were up, and she helped initiate Skywatch, an Ontario project monitoring pollution from the air.[14]

In 1984 Adele became the operations manager at Maple Airport, Ontario, and president of Neiltown Aviation. When the airport closed three years later she bought Aviation International in Guelph, Ontario. There she taught flying, chartered planes, did aircraft maintenance, sold fuel, offered sightseeing tours, and ran a restaurant. "My greatest rewards are seeing some of our graduates flying for the airlines or working in the industry," she says. Adele sold the company in 2011 after she'd been in the business for twenty-five years.

Adele has many interests. She has a brown belt in karate, which keeps her in excellent physical condition; she is a pianist and loves to "jam"; she enjoys gardening; and she is a first-rate photographer. In 1986 she was a member of the only all-woman team that participated in a balloon rally to celebrate Israel's fortieth anniversary. "I have the type of personality that will never say no to a challenge. When I see a door open, I run through it," she remarks.[15]

Margo Smith McCuthcheon (born 1938) also flew with Daphne and Adele. The three women came first in the women's section in a Transatlantic Air Race from New York to Paris in 1985. They were the only women's crew from Canada. During the flight they were chatting to each other when they realized they'd forgotten to turn their microphone off; every other pilot in the race could hear their conversation.[16]

(Left to right) Pilots Adele Fogle, Daphne Schiff, and Margo McCutcheon, 1985, just after they landed at Buttonville Airport (Toronto) at the end of their New York to Paris air race. *Courtesy Adele Fogle.*

When she was in high school Margo won a flight to New York from Toronto, but it wasn't until after she married and all her children were in school that she had time for lessons. "I was hooked from the first lift off.... I loved that summer, soaring high, banking and swooping in figure eights.... I love flying for the feeling of freedom, escapism from the real world and the need to be completely focused on something entirely different," she says.[17]

Margo has also been heavily involved in Skywatch operations (see chapter 15).

Recently Daphne and Adele have used their flying skills to help other people around the world. "I have a collection of cups [trophies] from all over that clutter up my place," Daphne says. She doesn't need any more. "We decided to do something more useful," she explained, so they focused on helping communities in Africa.[18]

For several years after 1998 they flew to Africa in a special non-profit program, Air Solidarité. The organization contributed the planes, but Daphne and Adele were required to raise at least $15,000 themselves for each flight. The money went toward projects in various countries, such as water purification plants, women's literacy programs, agricultural projects, health centres, schools, and solar panels for electricity. They've also helped set up a rehabilitation and prosthetic centre and a paramedical education project. Daphne and Adele canvassed Canadian companies for medicines and school supplies to take with them. Often where they visited they found the children didn't even have pencils or crayons.[19]

Each time Daphne and Adele flew to France, they picked up a small single-engine plane. They flew the plane themselves 3,730 miles to central Africa to deliver their supplies and visit other projects. They paid for all their own food and other expenses and the cost of fuel for the plane. Each of their trips lasted about a month. Daphne was flying even into her eighties. In fact, Daphne was eighty years old in 2004 when Adele had to opt out of a trip. Daphne flew by herself 1,864 miles in Africa in a single-engine airplane.

"In the southern hemisphere, the temperatures are opposite to those you experience at home," Daphne explains. "Over the Sahara [Desert], the contrast between day and night is extreme. Fuelling the aircraft was not easy." The fuel arrived in barrels delivered by trucks, "provided they were not hijacked. Then we pumped [the fuel] by hand to the top of the wing ... a two hour process in intense heat up to 40°C."[20]

The flights were often a challenge, she notes. Once when they were stranded in Libya for two days because there was no fuel, they had dinner with the Bedouins (the desert people) and "rented jeeps and drag-raced in the sand dunes." They've battled sandstorms and mosquitoes. Airports where they were to have landed have been closed because of governments' officials' visits, or airports exploding luggage that was suspect. They've been stranded by rain and fog, and sometimes visas for countries they intended to visit never arrived.[21]

But there were great rewards, too. "Whenever we landed, people lined the runways. The sight of children greeting the plane was overwhelming.... Some women walked fifteen miles in intense heat with babies on their backs to meet us ... And we've been handed roses — in Africa."[22]

"Flying gives you a chance to look at the world from God's point of view," Daphne says. "Even if it's a desert, it's shining and glowing. God did a pretty good job when She created the world. But when we land it's obvious that people aren't doing a very good job with the world God gave them."[23]

Daphne was given an honorary degree from York University in 2005. In her acceptance speech she spoke about sacrifice and giving back to society. "Never underestimate what you are capable of," she told the students. "Continue to follow your heart with your goal in mind. Think how you can make a contribution to better the world. Your reward comes from waking up every day, content with your life."[24]

Daphne established a graduate scholarship in psychology at York University, supported through donations from her friends and family.

15

Spies in the Sky: Operation Skywatch

In 1992 the president of a Canadian paint company was sent to jail for eight months. His company had stored liquid waste in barrels that leaked into the groundwater. Another company was fined $320,000.00 for polluting the earth.[1] Both charges were upheld in court because of photos taken from small Skywatch planes.

Operation Skywatch, begun in 1978, is a program run by the Ontario Ministry of the Environment with volunteer female pilots. For a few hours each month, a volunteer pilot flies a government inspector in the air who photographs places where there's been a complaint about pollution. This might be sewage lagoons, quarries, sewage treatment plants, landfill sites, or lakes.[2]

Some photos show rectangular shapes in fields that turn out to be rusting metal tanks that someone has buried, leaking toxic chemicals into the ground. Sometimes a path of dying plants and shrubs means that there are illegal pits full of harmful chemicals. Once a trucker was driving around the bottom of a gravel pit, deliberately draining his cargo of waste as he travelled, polluting the earth. He was photographed red-handed by the Skywatch planes.[3]

All Skywatch pilots must be members of the Ninety-Nines, an international organization of female pilots. The Ontario government provides the plane and pays all the costs, but the pilots volunteer their time. They're highly qualified: they must have flown 1,000 hours and have a commercial pilot's licence. This project allows women to fly, promotes women's role in aviation, and addresses environmental concerns.

Before each Skywatch flight the pilot finds out where the government inspector wants to go. She asks what pictures the inspector wants to take, looks at aerial maps, and then checks the weather. She needs to know if there are any towers, wires, or hills that she must avoid in the area. She has to check with airport control if she's flying near an airport. When she's in the air over the site she has to keep the plane steady so the photos will be clear.[4]

In 1994 Skywatch pilots flew over three hundred hours a year. In 1997 the pilots were in the air over Ontario about two and a half days a week in the summer, and about half that in the winter. The program was so successful that it was copied in the United States. By 2011, however, there were only four active pilots. Each pilot flew only ten to eighteen hours that year. The Ontario Government had cut the budget for the Skywatch program.

Akky Dillen Mansikka (born 1945) is one of the pilots who flies for the Skywatch program. In 1953, when she was eight years old, Akky came to Canada from the Netherlands by boat with her family. Because her parents and young brother were sick from the rough seas during the trip, Akky was able to wander about the ship on her own. She was fascinated by the "swirling pattern of the green, blue and black of the water" in the ocean.[5] Akky knew two English words, "yes" and "no," and thought that should be enough for her to get by in her new country.[6] She was excited by every detail of their trip. She watched Mickey Mouse movies and ate strange foods in the dining room, where she threw up on the table cloth.

"Every evening, the captain announced our position, latitude and longitude, which I marked on the back of the passenger list which had a map of the North Atlantic. Much to my dismay years later, I found I did not indicate the day and time of each position report. I ended up with just one continuous line representing our journey!"[7]

Akky says she'll never forget the October day they arrived in Halifax. There were "the most colourful trees I have ever seen — brilliant reds, yellows and oranges. I thought this must be a magic land."

The family immediately took a train to Toronto, or "cowboy country in the wild west," as her mother called it.

When she went to school she was put into grade one, even though she had been in grade three when they left their home. "I was the biggest kid and sat in the front row," Akky says. She felt out of place. "At recess the kids taught me words in English — not all of them polite!"[8]

Akky had always been interested in flying. In Holland there was an air force base across the canal from where she lived. After she came to Canada she would sometimes go with her father to the airport where he worked.[9] She took her first flying lesson in 1960 against her parents' wishes. "Women were not encouraged to pursue nontraditional careers," Akky said, so she became a teacher instead of a pilot. Akky took up flying lessons again when her kids were grown up. She realized that the time to pursue dreams may never come unless she took the steps to do them then.

Now Akky does Skywatch flights, helping to protect the land, water, and air we breathe. She points out that she still works with latitude and longitude, but she does a better job of plotting it than when she was eight. "This is a magic land, after all," she says, and she works to keep it that way.

Skywatch pilot Akky Dillen Mansikka.
Heather Bradacs.

In 2010 Akky took to the air to retrace the twenty-mile loop that the early flying ace, Jacques de Lesseps, flew over the city one hundred years earlier. Her flight was part of a program to honour the early birdmen. Akky grew up in the part of Toronto that once contained the airfield where Jacques took off for his spectacular flight, and came to realize that the city had been built around the old runway.[10]

"Flying makes your daily problems seem so minuscule," Akky says. "It brings faraway places closer to home." Her favourite places to fly to are the Bahamas, the Bruce Peninsula, and the north shore of Lake Superior.

Akky has lots of other interests as well. She hikes and officiates at downhill ski races and she's a dragon boat racer who has participated in championship races around the world.

16

Airplane Gymnastics: The Snowbirds

Female Canadian pilots have been doing "gymnastics" in the air, or aerobatics, since they first began flying planes. Eileen Vollick Hopkin, the first woman in Canada to become a licensed pilot, flew loops and rolls in 1928, the same year she received her pilot's licence.

Rolie Moore Barrett Pierce (1912–1999) was an aerobatic pilot called "an airport rat." She spent all her time at airports cleaning engines, washing airplanes, and sweeping the hangar in exchange for flying lessons. In 1936 her mother took her and her sister to England. Rolie stayed there to take aerobatic lessons. She won several trophies for flying in England and also when she came back to Canada.

In Canada, Rolie (a nickname for Rosalie) was hired by Associated Air Taxi in the 1950s. It was then that she began flying passengers up and down the west coast of British Columbia on float planes. "I flew from 8 a.m. to dark, and then some, and fit everything else around flying," she said. "I could always be sure of … getting wet feet."[1]

But doing stunts in an airplane the way Eileen and Rolie did alone, and being part of Canada's synchronized team, the Snowbirds, are quite different. When the nine Snowbirds fly in formation, they have to fly together perfectly. When they do the "Big Diamond" stunt, their airplane wings overlap. Only a few feet separate the planes. In the "Bomb

An F-86 Sabre aircraft flies with the Canadian Forces Snowbirds air demonstration team over Comox, B.C. The Sabre or "Hawk One" is refurbished and repainted in the golden colours of the RCAF Golden Hawks. *Royal Canadian Air Force Imagery Gallery, IS2009-0331, April 20, 2009. Photo by LCdr Kent Penney.*

Burst" all nine planes fly down together toward the ground, at the last moment splitting off in all directions. Their demonstrations have been described as a ballet in the sky.

"The job requires a steady 'three-dimensional mind,'" explains Maryse Carmichael, the first female Commander. "You have to know where everyone is ... [and] have an all-encompassing idea of what is going on around you."[2]

The Snowbirds began in 1970 with volunteer pilots, practising after their regular working hours. They were based in Moose Jaw, Saskatchewan where the elementary school held a contest to name the team. The Snowbirds won out, and in 1977 they became part of Canada's Armed Forces, now the 431st Air Demonstration Squadron.

It isn't easy to become a Snowbird pilot. Only four pilots are hired each year. The pilots can fly with the team for only two years because it's such a stressful and dangerous job; they have to have nerves of steel. The planes fly at a speed of more than 403 miles per hour. Seven pilots and one passenger have been killed since they began.[3]

The pilots get up at the crack of dawn every day to train. They have to learn new stunts every year, and half the year they're on the road, away from their families. Millions of Canadians and Americans watch them fly in air shows from May to October.

Maryse Carmichael (born 1971) from Beauport, Quebec, was five years old when she first saw the Snowbirds at the Bagotville Airshow. "I remember being mesmerized by those nine airplanes in the sky," she says. At that time the only woman associated with the aerial team worked behind a desk. Maryse couldn't imagine that one day she would be the first female pilot to fly one of those red and white jets. In fact, she'd become the first woman in charge of the whole team.[4]

When she was thirteen she became an Air Cadet, following her three older brothers. She started flying gliders when she was sixteen. By the time she was seventeen she was a licensed pilot, and in 1990, at age nineteen, she joined the Canadian Armed Forces, one of only two women in the flight school in a class of 140 people. "Receiving my Canadian Forces pilot wings in 1994 has to be my proudest moment," she recalls.[5] At one point in her career Maryse flew VIP transport around the world, carrying notables such as Prime Minister Jean Chrétien and Governor General Adrienne Clarkson.[6]

In 2001 Maryse became the first woman to fly with the Snowbirds and the first woman anywhere in the world to fly with a jet aerobatic team. She was in the number three position. The next year she flew in the number two position. She called it a dream come true.

Ten years later Maryse became the Snowbirds' first female commanding officer, in charge of eighty-five men and women, including fourteen pilots. She was promoted to lieutenant-colonel, based in Moose Jaw, Saskatchewan. "I was on the team a few years ago and I was the first woman to fly with the Snowbirds and that was really different. But now this posting is really about commanding the squadron and commanding the men and women that really represent the Canadian Forces across Canada." The position allows her to fly, but she no longer tours with the team.[7]

Lt. Col. (ret'd) Maryse Carmichael in her Snowbirds uniform, Toronto, 2010. *Ken Lin.*

"It's a fabulous challenge and a rewarding career," Maryse says. "When you're flying and you have a helmet on your head, it doesn't matter whether you're a man or a woman. As long as you can stay in formation, the public from the ground doesn't see it, [that you're a woman,] so that's the way I see it, too."[8]

Maryse is used to working with men and being equal to them. She grew up the youngest in a family of four and was the only girl. Now Maryse and her husband, fighter pilot Scott Greenough, share the work of bringing up their two young daughters, Georgia and Danielle. "It's the greatest thing [to share the work equally]," she admits. "There is no real competition between the two of us [because we're both pilots] … unless we're playing a car racing game."[9]

Maryse's advice to other women is to "move forward and don't be afraid. I just do what I love to do and I love flying and I'm fortunate enough that it has worked out." She tells both men and women "to follow their passion. If you love what you do, you will put in the hours and you will be dedicated to being the best you can be."[10]

In 2005 Maryse was named one of the Top One Hundred Most Powerful Women in Canada by the Women's Executive Network, and in 2013 she received the Northern Lights Award.

There have been other Canadian acrobatic teams, all male:

The Siskins, a team of three biplanes, flew from 1929 to 1932.

The Golden Hawks were formed in 1959 to celebrate the thirty-fifth anniversary of the Royal Canadian Air Force and the fiftieth anniversary of powered flight in Canada. A team of seven, they lasted until 1964 after performing 317 shows in their gilded Sabres. They always flew low over the spectators, opened their canopies, and waved.

The Golden Centennaires performed in 1967 for Canada's one-hundredth birthday. An eight-member team, they flew in blue, gold, and red jets, which the Snowbirds later used.

17

Over the Rainbow:
Female Astronauts

Two Canadian women have flown as astronauts, Roberta Bondar (born 1946) and Julie Payette (born 1963). They both are part of NASA, the National Aeronautics and Space Administration, in the United States.

Roberta Bondar always wanted to be an astronaut.

When she was seven years old she and her older sister Barbara made a cardboard and wooden spaceship with wire controls in the cockpit. They had Flash Gordon water pistols and listened to Commander Tom and Cadet Happy on the radio. They tried to contact beings from outer space on a radio set.[1]

When they were young, Roberta and Barbara saved all of their Double Bubble Gum wrappers and sent them in to the company in return for space helmets, but when the helmets came in the mail they were disappointing. Roberta had imagined something magnificent, but they came in a flat brown envelope. When the girls opened them up, the helmets were nothing but tall white cardboard rectangles with a large square opening at the front. While her sister built plastic models of ships and airplanes, Roberta built models of rockets, space stations, and satellites. When she was eleven she made a model of the first Soviet Sputnik satellite, launched that year. "I longed to soar into space. I wanted to reach out to adventure with my body as well as my imagination," Roberta says.[2]

She read science fiction comics and books about space. She imagined that she was in every rocket that was launched. She never wanted dolls. Instead she asked for science equipment, chemistry sets, a Brownie camera, a doctor's bag, and sports equipment, all of which she received for her birthdays and Christmas. Her family had a skating rink in their back yard where they lived in Sault Ste. Marie, and a laboratory with a microscope in the basement.

Roberta loved sports. She liked the rush she got when she competed in basketball and tennis; she liked canoeing and hiking. She biked, went hot-air ballooning, fished, skied cross-country, and played squash. When she was in university she coached archery and spent time on photography.[3]

She never gave up. When she didn't make the basketball team in grade nine her sister coached her and she made the senior team the next year. When she was in grade thirteen the guidance counsellor tried to talk her out of studying science because she was a girl, but she did a science project on the moon and even became the leader of the school science club.

By the time she was twenty-two Roberta had a university degree in science and a pilot's licence, but she wanted to be a doctor, too. The first time she applied to medical school she was turned down because her marks weren't good enough, but that didn't stop her.[4] She did scientific research on the heart, earning a master's degree. Then she studied the fish's central nervous system for her PhD. The next time she applied to medical school she was successful. She graduated as a doctor when she was thirty-one years old.

Roberta was teaching medicine at McMaster University and directing a clinic for multiple sclerosis (MS) when NASA advertised needing men and women to fly as astronauts. In 1982 Canadians were invited to apply, too. Roberta applied, but so did more than 4,000 other men and women. When Roberta was chosen as one of only nineteen astronaut finalists, and the only woman, her mother gave her the rocket she had made as a child. She had saved it all those years.

On January 22, 1992, Roberta blasted into space, the second Canadian astronaut and the first Canadian woman. Sixty-five of her friends and

Astronaut Roberta Bondar.
Photo by CarlyleRouth.

relatives waited in a private viewing space. As she walked towards the shuttle *Discovery* she raised both hands in the air and called out, "Yes, yes, yes," to all those watching.[5]

For four hours she lay on her back strapped into place, wearing ninety pounds of special clothes for the launch and re-entry of the rocket. There was a deafening roar. The whole spaceship shook as it began the journey into space. Roberta says it felt like riding a monster roller coaster with a gorilla sitting on her chest. Then suddenly she was weightless. Roberta was living her childhood dream.

Roberta and six other members of the crew went around the earth 129 times, alternating forty-five minutes of light and forty-five minutes of darkness. They crossed North America in ten minutes each time they orbited the earth. They were away for 193 hours — over eight days.[6]

Roberta worked as a payload specialist on the trip — that is, she was in charge of several scientific experiments. She watched to see how plants would grow in space; she studied the effects on people in space; she worked on space sickness; she watched as the people in her crew got taller in space and had better vision, but when they returned to earth, they shrank again and had to wear their glasses. There were forty-two experiments on that flight, and Roberta enjoyed her tasks. She wanted to be known "as a highly trained scientist, not someone who was selected just because she was a woman."[7]

Roberta left the space agency after her flight. Fascinated by the views of Earth from space, she became a professional photographer, and has since photographed all of Canada's national parks. She also teaches and writes.

When Roberta was young she believed she could do anything, and that her life would be an adventure. It has been. She tells boys and girls to "get as much education as you can." She was determined to get every "piece of paper" or university degree she could because she wanted every door to be open for her; she wanted to be able to do whatever she decided.

"You need to have confidence," she says. "As long as you have the skill set, it's all about confidence…. Success deals with not being arrogant, but rather about being confident."[8]

Julie Payette was the first Canadian to work in the International Space Station. In 1992 she was chosen as one of only four new astronauts out of more than five thousand men and women who applied. She was flabbergasted.

When Julie was nine years old she decided to become an astronaut. She watched the Apollo space mission to the moon when she was in school in Montreal, where she lived. She'd never even been in an airplane, but suddenly she knew what she wanted to be. She made scrapbooks of space flights and had pictures of astronauts on her bedroom door. "One day I'll make an enormous pop right into orbit around the earth and contemplate the world," she wrote in her school yearbook.[9]

"I was very lucky to be in a family that didn't just laugh at me," Julie says. When her mom and dad saw that she was serious, they said "OK, well you want to do that? Well, you better work, you better go to school, you better be good." She had a physics teacher who encouraged her to study science.[10]

Julie decided that a first step would be to study at university to become an engineer. She believes that being an engineer teaches you to look at problems, to learn how to solve them, and to design something different. "That's exactly what we do in space. You need to be able to repair things and design new things, and you certainly have to … look at a problem and try to solve it," she explains.

"I was a good student," she says, "but I was not at the top of my class."[11] But Julie became well rounded; she didn't just study science at school. She learned to play competitive handball and badminton; she runs, skis, hikes, and scuba dives; she's done triathlons; she learned to play the piano; she sung in several well-known choirs; she speaks French, English, Spanish, and Italian, and she learned Russian to work in the space program. She also became a pilot, flying a range of aircraft, from jets to float planes.[12]

It's important to be able to do a lot of things, Julie says. When you're in space, you have to be a Jack of all trades. She points out that in a space ship "we're the only ones on board. We're the repairman, we're the cook, we're the photographer, we're the document keeper, we're the cleaner, we're the … scientist … the spacewalker. We … do everything. So it's not about being the top in one field," she notes, "it's about being good in many areas."[13]

It's also important to be part of a team, Julie explains. She believes that singing in a choir was very good training. When you sing in a choir, you have to work as part of a group; you need to all sing the same tune.

Astronaut Julie Payette.
Canadian Space Agency.

Once Julie was selected to become an astronaut, she spent years training to go into space. She had to learn what to do in case there was an emergency. Space is very hard on people and equipment. "It's scorching hot when you're exposed to the sun, and extremely cold when you're not in the sun, and it varies every forty-five minutes, from -150 degrees Celsius to +150 degrees Celsius). So you can imagine what it does to people or to equipment if ... [your space journey] is not well planned ... There's weightlessness. There are radiation doses that are much higher outside the atmosphere of the Earth," she adds.[14]

On May 27, 1999 Julie and six other crew members burst into space for ten days onboard the space shuttle *Discovery*. Forty hours later they docked at the International Space Station. They took four tons of equipment and supplies with them so that astronauts after them could live at the station. Julie supervised an eight-hour spacewalk to repair and add to the station, and also operated the Canadarm.

Julie was back in space a second time in 2009 onboard the space shuttle *Endeavour* — this time for fifteen days. Now she was the flight engineer (part of the flight crew responsible for all vehicle manoeuvres, including launch, re-entry, and docking at the space station) and the lead robotics officer, operating three different robotic arms. The *Endeavour* took more equipment for the station.[15] The first time Julie visited the space station it was small. The second time she went there it was the size of a three-bedroom house with six people living in it.

Julie knows that in the next decades someone will fly to Mars. She probably won't go there, but she loves being in space. She quotes the great Italian artist and scientist Leonardo da Vinci (1452–1519), who lived centuries ago before airplanes were even invented but was arguably the first to envision flying machines: "... once you have tasted flight, you will forever walk the earth with your eyes turned skyward, for there you have been and there you will always long to return." That's how Julie feels about space.

"I will say to my children: choose your path. Don't choose somebody else's path. Look at what you are, what you like, what your interests are, what you are good at, because we are all good at several things, and dare to dream!" Julie advises.[16]

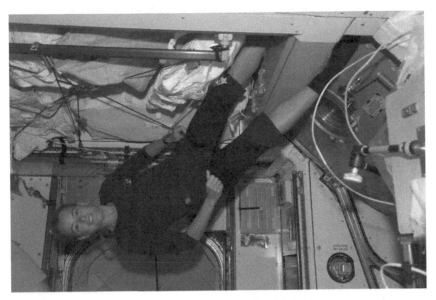

Astronaut Julie Payette floating in her space capsule, July 20, 2009.
NASA.

Roberta and Julie have flown further into space than any other Canadian woman. A hundred years ago the first female airline passengers wouldn't have believed these space trips were possible. But space exploration is likely just in its beginning stages. Someday soon people will explore the planet Mars. Perhaps Canadian women will be on the first space ship that travels there.

From 1963 to 2015 fifty-eight women from around the world travelled into space as astronauts. Valentina Tereshkova from Russia was the first woman to travel in space on June 16, 1963, in *Vostok 6*, flying for forty-eight orbits — a total of seventy hours and fifty minutes.

Afterword

In 1924 the tagline for a Palmolive soap advertisement ran: "Most men ask 'Is she pretty?' not 'Is she clever?'" In 1953 an ad for Del Monte ketchup bottles with screw tops read: "You mean a woman can open it?"[1] Whether or not one or both copywriters were being tongue-in-cheek, both ads reflect societal stereotypes that were all too prevalent — stereotypes that worked against the most talented and conscientious women who sought to carve out careers as airplane pilots. If a woman did not need to be clever and was not mechanically minded enough to open a screw-top bottle, how could she possibly fly an airplane?

This thinking persisted even into the last decade. Johanna Flygare from the School of Aviation in Lund University, Sweden, notes that the role of the flight attendant, generally associated with women, is a caring role, while the pilot tends to be "a male dominated top of the hierarchy occupation.... What proves leadership, management and masculinity more than controlling and leading a big machine like an airplane?" she asks.[2]

Even today, eHarmony dating advice on the internet reports that male pilots are the third most attractive males to women, while a female flight attendant is the sixth most popular female to men. One would need to examine these numbers scientifically and analytically before coming to a conclusion, but perhaps those men who are looking for a female flight attendant are seeking a woman with a "caring" personality and not a "controlling" or "leadership" one. Both male pilots and female flight attendants receive the second most messages from the opposite sex according to Dr. Steve Carter, eHarmony's Vice President of Matching.[3]

In spite of such attitudes, which we have seen demonstrated in the preceding stories, many passionate and determined early female flyers did achieve their goals, in many instances supported by both men and women. But a number of the women also achieved their goals, only to abandon their career when they married.[4] The socialite Grace Mackenzie appears to have settled down to a supportive role suitable for her time, and the adventurous Madge Graham apparently had no great desire to become a pilot, but only to bring up her children, later describing her flights as "stupid" but "fantastic." The pioneer pilot Eileen Vollick Hopkin abandoned her career after she married and had children, Enid Norquay McDonald stopped flying because her husband disapproved, and Louise Mitchell Jenkins was persuaded by her friends to give up her great passion when her husband fell ill.

As one of the female pilots explained, flying is an expensive hobby. Unless a woman has her own independent income, even today, she is dependent on her partner or someone else for financial and personal support, and some partners may not approve of such a career, or even a career outside the home at all.

It is interesting to note that some early female pilots from the United States were welcomed by Canadian society even as Canadian women were discouraged from flying. One can suggest reasons for this discrepancy. Not only did the United States have a greater population to choose from, but in the nineteenth and early twentieth centuries women in that country had greater opportunities and support because, on the whole, it was a more progressive and democratic society. Women from the United States had a reputation of being more independent than Canadian women, who were seen as being cast into a more submissive mould.[5] A nineteenth-century American farmer is reported to have said that he would go to Canada for a wife when he married. "When I come home at night she'll have a nice blazing fire on, and a clean kitchen, and a comfortable supper for me; but if I marry a New Yorker, it'll be when I come home, 'John, go down to the well for some water to make the tea;' or 'John go and bring some logs to put on the fire to boil the kettle.' No, no, a Canadian woman's the wife for me."[6]

It is also interesting to note that the few female balloonists whose stories opened this book were warmly welcomed here in Canada. In fact their daring exploits were the highlight for celebrations. Perhaps a woman in a balloon was not as threatening to society as a woman in a more mechanical and industrial-looking machine. But then, most of the balloonists, or "aeronauts," as they were called, who performed in Canada —at least those we're aware of — were from the United States or Europe.

Nevertheless, as the preceding stories show, in spite of all the strictures against female pilots, women in Canada have made a tremendous contribution to the aviation industry over the past one hundred years, and continue to do so.

APPENDIX
Milestones in Aviation

1903	Americans Wilbur and Orville Wright fly at Kitty Hawk, North Carolina
1909	Dolena MacKay MacLeod, first Canadian woman to fly as a passenger
1910	Baroness Raymond de Laroche of France, first licensed female pilot in the world
1911	American Harriet Quimby, first licensed female pilot in the United States
1913	American Alys McKey Bryant, first female to fly a plane in Canada
1919	Marguerite Marie "Madge" Vermeren Graham, first female Canadian airplane navigator
March 1928	Eileen Vollick Hopkin, first licensed female pilot in Ontario and in Canada
Oct. 1928	Eileen Magill Cera, first licensed female pilot in Manitoba
Dec. 1928	Gertrude de La Verne Tanner, first licensed female pilot in Alberta

Aug. 1929   Daphne Paterson Shelfoon, first licensed female pilot in New Brunswick

Oct. 1929   Nellie Carson, first licensed female pilot in Saskatchewan

Nov. 1929   Jeanne Grenier Gilbert Fothergill, first licensed female pilot in British Columbia

1931   Peggy (Ethel) Ricard Standring, first licensed female pilot in Nova Scotia

1932   Louise Mitchell Jenkins, first licensed female pilot in Prince Edward Island and first Canadian woman to buy an airplane

1932   American Amelia Earhart, first woman to fly solo across the Atlantic Ocean

1936   The Flying Seven organized in British Columbia

1938   Lucile Garner Grant and Pat Eccleston Maxwell, the first two stewardesses hired in Canada

1939   Elizabeth (Elsie) MacGill Soulsby, the first woman in the world to design a plane that flew

1941   Jean Flatt Davey, the first female doctor to enlist in the Royal Canadian Air Force, Women's Division

1943   Elspeth Russell Burnett, first licensed female pilot in Quebec

1947   Violet Milstead Warren, first female bush pilot.

1947   Gertrude Dugal, first francophone female pilot in Quebec

1947   Phyllis Penney-Gaul, first licensed female pilot in Newfoundland

1949   Marion Alice Powell Orr, first Canadian woman to own and operate a flying school and an airport

1959   Molly Moretta Fenton Beall Reilly, first female Canadian captain, flew with Canadian Utilities Company

1973   Rosella Bjornson, first Canadian woman hired by a major commercial airline, Transair

1974   Major Wendy Arlene Clay M.D., the first woman to receive
             military wings in Canada

1978   Corporal Gail Toupin, first female member of the Sky Hawks,
             the Canadian Army's parachute skydiving team

1978   Judy Evans Cameron, first female pilot with Air Canada

1978   Skywatch begins

1979   Dee Brasseur, Leah Mosher, Nora Bottomley, first female pilots
             with the Canadian Armed Forces

1989   Dee Brasseur and Jane Foster, first female Canadian fighter
             pilots

1990   Rosella Bjornson, first female pilot and captain with Canadian
             Airlines International

1992   Corporal Marlene Shillingford, first female technician with the
             Snowbirds

1992   Roberta Bondar, first female Canadian astronaut

2000   Lieutenant-Colonel Maryse Carmichael, first woman to fly with
             the Snowbirds

2009   Julie Payette, first female, Canadian astronaut to enter the
             International Space Station

2010   Lieutenant-Colonel Maryse Carmichael, first female
             Commanding Officer of the Snowbirds

NOTES

Preface

1. Emine Saner, "Female Pilots: A Slow Take-Off," *Guardian*, January 13, 2014.
2. Statistics Canada, 2011.

Introduction

1. Rowena Davison and Reuben Davison, *An Edwardian Housewife's Companion: A Guide for the Perfect Home* (Newbury Park, California: 2009), 20.
2. *Canadian Aviation Historical Society* 5, no. 3 (Fall 1967), 68.
3. Ken Tigley, "Historical Impact Assessment," June 2009.

BEFORE THERE WERE PLANES

Chapter 1: Up, Up in a Basket

1. "The Almonte Celebration," *Almonte Gazette*, September 5, 1879.
2. "Quebec's First Lift-Off was in a Balloon Named Canada," *Quebec Heritage News* 2, no. 11 (July 2004), 7; "A Sleeper of Sorts," *The Canadian Aerophilatelist*, December 2002, 6; Renald Fortier, "The Balloon Era," *Canada Museum*, 2004.
3. Donald Dale Jackson, *The Aeronauts* (Alexandria, Virginia: Time-Life Books, 1980), 96.

4. "Grand Balloon Ascension at McFarlane's Grove in 1879," *Almonte Gazette*, reprinted July 30, 1970.

5. "The Almonte Celebration, an Immense Success," *Almonte Gazette*, September 5, 1879.

6. "Grand Balloon Ascension at McFarlane's Grove in 1879," *Almonte Gazette*, reprinted July 30, 1970.

7. F.H. Hitchens, "Ballooning was safer in Canada," *CAHS Journal*, Summer 1967, 45.

8. "Odds and Ends," *Otago Witness*, no. 1464, December 6, 1879, 6.

9. Gordon Beld, *The Early Days of Aviation in Grand Rapids* (The History Press, 2012), 21; F.H. Hitchins, "Ballooning was safer in Canada," *CAHS Journal*, Summer 1962, 45; Doug Grant, "Aeronaut Prof. Squires [sic] Provides an Exciting Day in Brockville — 1874," *Handbook of Brockville History*.

10. Hitchins, "Ballooning was Safer in Canada," 45.

11. "Grand Balloon Ascension at McFarlane's Grove in 1879," *Almonte Gazette*, reprinted July 30, 1970; "Nellie C. Moss Squire," Find a Grave, www.findagrave.com. Her tombstone notes that she was born in 1859, although her obituary states 1846; Hobie Morris, "Central New York's First Aerial Heroine: Balloonist Nellie Thurston Squire," *Musings of a Simple Country Man*, n.d.; "Today in Aviation History," Pilots of America, www.pilotsofamerica.com; Ken Tigley, "Historical Impact Assessment," June 2009. Many years later, in the 1950s, women were parachuting — but for a good cause. Para-rescue nurses, called "para-belles," were parachuting into remote areas of Canada on rescue missions as part of the military's search and rescue operations. The most famous was Grace MacEachern, the first woman "para-belle," whose first jump on Mount Coquitlam in British Columbia landed her in a tree, taking her about two hours to get down; *On Windswept Heights* (The Royal Canadian Air Force, 2013), 52.

THE EARLY YEARS

Chapter 2: Daredevil Female Passengers

1. Verified in *Cape Breton Post*, 1959; also in Parks Canada, "Aerial Experiment Association Chronology," Alexander Graham Bell National Historic Site; George Fuller, "Free Flight," n.d.; Harold A Skaarup, *Canadian Warplanes* (Bloomington, Indiana: IUniverse, 2009), 4.

2.   As described by Frank Ellis in "Pioneer Flying in British Columbia, 1910–1914," 227.

3.   George Fuller, "The First Canadian Woman to Fly," *CAHS Journal*, Summer 1969, 56. This source suggests that the usual date for this flight, July 1909, was in error.

4.   JA Douglas McCurdy was the first person to fly in Canada, on February 23, 1909.

5.   Sean Chase, "The Extraordinary Exploits of Casey Baldwin," *Daily Observer*, November 8, 2006.

6.   While these three men were the actual inventors, it was Alexander Graham Bell's wife and student, Mabel Hubbard Bell (1857–1923), deaf since she was five years old as a result of having suffered scarlet fever, who pulled the whole operation of the Aerial Experiment Association together, and who kept tabs on their finances. At the Association's last meeting, JA Douglas McCurdy read a special resolution into the minutes: "Whereas members of the Aeronautical Experimental Association individually and collectively feel that Mrs. Alexander Graham Bell has, by her great personal support and inspiring ideas, contributed very materially to any success that the Association may have attained, resolved that we place on record our highest appreciation of her loving and sympathetic devotion without which the work of the Association would have come to naught ..." "Mabel Bell," *Section 15*, http://section15.ca/features/people.

7.   Sean Chase, "The incomparable exploits of Count Jacques de Lesseps," *Daily Observer*, October 14, 2010.

8.   "Successful Flights at Montreal Meet," *Knoxville Journal*, June 29, 1910; Frank Tibbo "(Long) live the count," *Gander Beacon*, July 14, 2008.

9.   Sean Chase, "The Incomparable Exploits of Count Jacques de Lesseps," *Daily Observer*, October 14, 2010.

10.   Carl Mills to Ralph Cooper, *Biographical Notes*, July 24, 1909.

11.   "The Lafayette Negative Archive," New York City; *Times*, January 26, 1911, 11.

12.   "de Lesseps, Tauni," obituary, *New York Times*, March 29, 2001.

13.   Sean Chase, *Daily Observer*, October 14, 2010.

14.   To the memory of Jacques de Lesseps, Knight of the Legion of Honour, Cross of War, Distinguished Service Cross, U.S. Born in Paris, 1883. The second to cross the English Channel in an airplane, 1910. The first

flight over Montreal and Toronto 1914–1918. Four times cited for the Ordre du Jour. Lost at sea Oct 18, 1927 while making aerial maps of the Gaspé coast. Cited by the French Nation Order. His body rests in the Gaspé cemetery.

15. The first flight in British Columbia had taken place on March 25, 1910, with the pilot Charles K. Hamilton. That year a race between a horse and an airplane was won by the horse. Frank H. Ellis, "Pioneer Flying in British Columbia, 1910–1914," 228, 230.

16. Ellis, "Pioneer Flying in British Columbia, 1910–1914," 242.

17. "The History of Flight in British Columbia," Canadian Museum of Flight, 2013. *Daily Province*, April 25, 1912.

18. Ellis, "Pioneer Flying in British Columbia, 1910–1914," 245; Joyce Spring, "The Sky's The Limit," *Canadian Women Bush Pilots* (Toronto: Natural Heritage Books, 2006).

19. "The Magnificent Distances, Early Aviation in British Columbia," Sound Heritage Series, Canadian Museum of Flight, n.d.

Chapter 3: The Flying Schoolgirl: Katherine Stinson

1. John H. Lienhard, "No. 1251: Katherine Stinson," *Engines of Our Ingenuity*, n.d.

2. Tony Cashman, "Katherine Stinson," U.S. Centennial of Flight Commission, n.d.

3. "WAI Pioneer Hall of Fame," Women in Aviation International, www.wai. org/pioneers.

4. Frank H. Ellis, "Katherine Flies the Mail," *CAHS Journal* 12, no.1 (Spring 1974), 12.

5. Ibid., 12–14.

6. "Bill Stout Meets Katherine Stinson," *Collection of Sumner B. Morgan*.

7. Ellis, "Katherine Flies the Mail," 12–14.

8. Debra L. Winegarten, *The Flying Schoolgirl* (Fort Worth: Eakin Press, 2000).

9. Some sources report that there were 249 letters.

10. Alan K. Spiller, "Katherine Stinson in Canada," 2007.

11. "You've Got Mail … an Alabaman Aviatrix in Alberta," *Vintage Wings of Canada*, www.vintagewings.ca/Vintage News/Stories/tabid/116/.

12. Tony Cashman, "The Katherine Stinson Special," *CAHS Journal* 44, no.1, (Spring 2006), 4–11.

13. J.R. Ellis, "Katherine Stinson in Ontario," *CAHS Journal* 9, no.1 (Spring 1971).

Chapter 4: It's a Bird ... It's a Plane ... It's Madge Graham

1.   Sylvia Graham, personal electronic correspondence with author, October 17, 2011; Sandra Jenson, "Learning to Fly," *Stories of the Stuart Graham Archives*, Concordia University, 2010. Notes that the hospital was in Manston, England.
2.   "Arrives Here by Bird Route from Halifax," *Daily Telegraph* (St. John), June 6, 1919.
3.   Ibid.
4.   Ibid.
5.   Bruce Ricketts, "Madge Graham, Canada's First Lady of Flight," *Mysteries of Canada*.
6.   Batty Baird, "Treetop Airwoman," John McInnes and Emily Hearn, *Sleeping Bags and Flying Machines* (Don Mills, Ontario: Thomas Nelson & Sons, 1973).
7.   Bruce Masterman, "Bush Pilots," *West*, Spring 2005.
8.   Stuart Graham, "The Beginning of Bush Flying," *Canadian Aviation Historical Journal*, Winter 2002, 124.
9.   *Canadian Forestry Journal* 15, no. 5 (June 1919).
10. "Century of Flight, Canadian Bush Pilots," *CBC News in Review*, February 2004, 56.
11. "People Flying Boat Patrols," Canadian Bushplane Heritage Centre.
12. According to a telephone conversation with her granddaughter.
13. Batty Baird, "treetop airwoman," John McInnes and Emily Hearn, *Sleeping Bags and Flying Machines* (Thomas Nelson & Sons [Canada], 1973).

Chapter 5: The American Influence

1.   "The Wright Brothers First Flight, 1903," EyewitnesstoHistory.com, 2003.
2.   "WAI Pioneer Hall of Fame," Women in Aviation International, www.wai. org/pioneers.
3.   "The American Girl Whom All Europe Is Watching," *World Magazine*, April 11, 1909.
4.   John Magee, "Great Aviation Quotes: High Flight."
5.   "Miss Quimby wins air pilot license," *New York Times*, August 2, 1911, 7; "Miss Quimby dies in airship fall," July 2, 1912, 1; "Harriet Quimby," National Aviation Hall of Fame; "The Remarkable Harriet Quimby," www.todayifoundout.com.
6.   Harriet Quimby, "A Woman's Exciting Ride in a Racing Motor-Car," *Leslie's Weekly*, October 4, 1900.

7. "Lafayette Girl Famous Aviator," *Courier*, March 1, 1917.

8. 2000 National Air and Space Museum, Smithsonian Institute.

9. D. Cochrane and P. Ramirez, "Alys McKey Bryant," Smithsonian National Air and Space Museum; Frank H. Ellis, "Pioneer Flying in British Columbia 1910–1914," 254.

10. Mrs. John Milton Bryant to unknown, 1912.

11. Ellis, "Flying in British Columbia 1910–1914," 255.

12. "The Magnificent Distances, Early Aviation ion British Columbia," Sound Heritage Series, Provincial Archives of British Columbia.

13. "The Alys Bryant Collection," Wichita State University.

14. Alys Bryant to unknown, 1912.

15. "Amelia Earhart," Women in Aviation and Space History, Smithsonian National Air and Space Museum.

16. Amelia Earhart, *The Fun of It*, as quoted in "Great Aviation Quotes."

17. Louise Aird, "High Flyers in History: Women in Aviation," *Arts Alive*, March 1998.

18. "WAI Pioneer Hall of Fame," Women in Aviation International, www.wai. org/pioneers.

DREAMS CAN COME TRUE

Chapter 6: Early Pioneers

1. *Maclean's*, reproduced covers from June 1930 and March 1, 1930, October 15, 1955, 29.

2. "Masculinity and the Making of Trans-Canada Airlines, 1938–1940," Spring 1994, 7.

3. E.M. Vollick, "How I Became Canada's First Licensed Woman Pilot," *Owen Sound Sun Times*, August 6, 2008.

4. Ibid.

5. Marilyn Dickson, "Eileen Vollick," Women Aviation Pioneers, *First Canadian 99s*.

6. Vollick, "How I Became Canada's First Licensed Woman Pilot."

7. *Bracebridge Gazette*, June 1, 1933, as reported in *Maclean's Magazine*, October 15, 1955, 101; Kelly Putter "Trailblazing the Airways," *Wings*, May 4, 2009.

8. Dickson, "Eileen Vollick"; "Gets Pilot's Licence," *Evening Telegram*, March 15, 1928, 27.

9. Dickson, "Eileen Vollick." Jacqueline denied her family connection,

although she kept in touch with them. "WAI Pioneer Hall of Fame," Women in Aviation, www.wai.org/pioneers.

10. "Aviation Pioneer, Eileen Vollick honoured," *Wiarton Echo*, August 6, 2008. The stamp was designed by Suzanne Wiltshire.

11. "Enid McDonald," obituary, *Guadalajara Reporter*, June 23, 2006.

12. Ibid.

13. Shirley Render, *No Place for a Lady, The Story of Canadian Women Pilots 1928–1992* (Winnipeg: Portage and Main Press, 1992), 31.

14. Ricketts, "Madge Graham, Canada's First Lady of Flight." Presentation to Island Heritage Committee, February 23, 2008; Syd Clay, "An Aerial June Wedding," Canadian Aviation Historical Society, PEI Branch.

15. Ross Smyth, *The Lindbergh of Canada, The Erroll Boyd Story* (Renfrew, Ontario: General Store Publishing House, 1997).

16. Ibid.

17. "Carl F. Burke MBE," "Famous Doctor Jack and Louise Jenkins of PEI and their aviation prowess," *CAHS Journal* 19 (Prince Edward Island: June 2006).

18. Render, *No Place for a Lady, The Story of Canadian Women Pilots 1928–1992*, 34.

19. Ross Smyth, *The Lindbergh of Canada, The Erroll Boyd Story.*

20. Shirley Allen, "Marion Alice Powell," *Pioneer Women Pilots*, Flashback Canada.

21. Lisa Wojna, *Great Canadian Women* (Edmonton: Folklore Publishing, 2005).

22. Patrick Watson, "Marion Alice Orr," *The Canadians: Biographies of a Nation* vol. 3, 2002.

23. Render, *No Place for a Lady, The Story of Canadian Women Pilots 1928–1992*, 141.

24. Bruce Forsythe, "Vaughan Perspectives in Canadian Military History," August 30, 2008.

25. Wojna, *Great Canadian Women.*

26. Vera Dowling, "The Amazing Vera Dowling Story," Max Solbrekken World Mission.

27. Render, *No Place for a Lady, The Story of Canadian Women Pilots 1928–1992*, 106.

28. Warren Hathaway, *Pursuit of a Dream: The Story of Pilot Vera Strodl Dowling* (Warren Hathaway, 2012), 6.

29. Ibid., 166.

30. Ibid.,106.

31. Ibid., 106, 135.

32. Dowling, "The Amazing Vera Dowling Story."

33. Hathaway, *Pursuit of a Dream: The Story of Pilot Vera Strodl Dowling*, 3.

34. Ibid., 4.

35. Ibid.; Dowling, "The Amazing Vera Dowling Story,"

36. Ibid.

37. Dorothy Rungeling, "Felicity McKendry — She had a Dream," Kingston Flying Club, October 20, 2002.

38. Ibid.

39. "Felicity McKendry," *The Canadian 99s*.

40. Render, *No Place for a Lady: The Story of Canadian Women Pilots 1928– 1992*, 180.

41. Felicity McKendry, *I Grew Too Tall* (Ottawa: Felicity McKendry, 2014), 9.

42. Ibid., 10.

43. Felicity McKendry, email correspondence with author, June 24, 2014.

44. Rungeling, "Felicity McKendry — She had a Dream"; Felicity McKendry (Ottawa: Felicity McKendry, 2014), 1.

45. Brian Johnson, "Flying Solo: The Felicity McKendry Story," April 13, 2013.

46. Felicity McKendry, telephone correspondence with author, June 13, 2014.

47. Rungeling, "Felicity McKendry — She had a Dream,".

48. McKendry, "Letter to the editor," *Kingston Whig-Standard*, September 27, 2012.

49. McKendry, telephone correspondence with author, June 13, 2014.

50. McKendry, email correspondence with author, June 27, 2014.

51. "Northern Lights Award — 2013 Award Recipients," Centennial College, www.northernlightsaward.ca/2013.

52. Brian Johnston, "It's Been a Magical Ride," *Kingston Whig Standard*, April 29, 2010, 129; Marilyn Dickson, ed., *Canadian Ninety-Nines 2006 Here and Now* (Toronto: Margo McCutcheon, 2007), 56.

53. Kristine Archer, "The Sky Is Not the Limit," *Vanguard*, July 2006.

54. "Rosa-One-One," *Speaking of Flying* (San Clemente, California: The Aviation Speakers' Bureau, 2000), 205; Archer, "The Sky is not the Limit."

55. Patricia Lovett-Reid, *Get Real* (Toronto: Key Porter Books, 2007), 253.

Chapter 7: Barbara Ann Scott: Queen of the Blades (1928–2012)
1.   "Skating Queen Likes Flying," *Canadian Aviation*, April 1947, 95.
2.   Barbara Ann Scott King, telephone correspondence with author, July 29, 2011.
3.   Ibid.
4.   "Queen of the Blades," *Between Ourselves*, Trans-Canada Airlines, February 1947.

Chapter 8: In the Captain's Seat: Rosella Bjornson
1.   Render, *No Place for a Lady, The Story of Canadian Women Pilots 1928–1992*, 287.
2.   Ibid.
3.   Ibid., 288; Emily Howell Warner became the first permanent woman pilot for a scheduled U.S. passenger airline in January 1973, for Frontier Airlines, and the first woman airline captain in 1976. "Emily Howell Warner," Women in Aviation International, www.wai.org/pioneers.
4.   "Rosella Bjornson," Library and Archives Canada, http://epe.lac-bac. gc.ca/100/200/300/ic/can_digital_collections/high_flyers.
5.   John Chalmers, "41st Annual Induction," *The Flyer* 32, no. 1, May 29, 2014.
6.   "Rosella Bjornson, Canadian Hall of Fame Inductee," *Canadian Flight*, November 1999.
7.   The stamp was designed by Suzanne Wiltshire.
8.   Gillian Holmes, ed., *Who's Who of Canadian Women 1999–2002* (Toronto: University of Toronto Press, 1999).

DURING THE SECOND WORLD WAR
Chapter 9: Angels in the Clouds: Early Stewardesses
1.   Richard J. Léger, "The Air Hostess," National Aviation Museum.
2.   Carol Toller, "The Sky's The Limit," *Chicago Tribune*, November 5, 1995.
3.   Newby, Jill, *The Sky's the Limit* (Vancouver: Mitchell Press, 1986), 1. Albert Hofe was hired by Deutsche Lufthansa in 1928, the first airline steward. Twenty-five-year-old Ellen Church was hired by United Airlines in 1930. Anthony J. Stanonis, "Femininity in Flight Review," *Canadian Journal of History*, 2008.
4.   "Lucile Garner was Canada's First Airline Stewardess," obituary, *Globe and Mail*, March 28, 2013.
5.   Ibid.; NetLetter, Terry Baker, n.d.; Peter Pigott, *Air Canada: The History* (Toronto: Dundurn Press, 2014), 301.

6. "Stewardesses of the 1950s," Library and Archives Canada, http://epe.lac-bac.gc.ca.
7. Newby, *The Sky's the Limit*, 108; Wendy Travis, "On the Wings of a Nightingale," spring 1986.
8. Ann Brown, hired as a stewardess by United Airlines in 1940, is credited with forming the first stewardess union at United, the Air Line Stewardess Association, in 1945. "Ann Brown," Women in Aviation International, www.wai.org/pioneers.
9. Jill. *The Sky's the Limit*, 1; Albert J. Mills and Jean Helm Mills, "Masculinity and the Making of Trans-Canada Air Lines 1938–1940: A Feminist Poststructuralist Account," *Canadian Journal of Administrative Sciences*, 2006, 12.
10. "Lucile Garner was Canada's first airline stewardess," obituary, *Globe and Mail*, March 28, 2013.
11. "TCA's First Stewardess, Lucile Garner Grant," *NetLetter for Air Canada's Retirees*, no. 1126, June 26, 2010.
12. "TCA Girls Hold Reunion," NetLetter no. 1239, *Horizons*, July 1973.
13. "Stewardess Tells Odd Tale," *Ottawa Evening Citizen*, January 20, 1948.
14. Richard J. Léger, "The Air Hostess," National Aviation Museum.
15. "Lucile Garner was Canada's First Airline Stewardess," obituary, *Globe and Mail*, March 28, 2013; NetLetter, Terry Baker, n.d.
16. Danielle Metcalfe-Chenail. *Polar Winds: A Century of Flying the North* (Toronto: Dundurn Press, 2014), 113.

Chapter 10: Margaret Fane Rutledge and the Flying Seven

1. "The History of Metropolitan Vancouver," *Vancouver History*, www.Vancouverhistory.ca/archives_flyingseven.htm.
2. Tom Hawthorn, "Margaret Fane Rutledge," *Globe and Mail*, January 5, 2005.
3. In 1941 the RCAF began employing women in a new branch — the Canadian Women's Auxiliary Air Force — to release men "so that men may fly." Over the course of the war the RCAF hired more than 17,000 women, first in nine trades: clerks, cooks, equipment assistants, fabric workers, general duties, hospital assistants, motor transport, drivers, and telephone operators, and soon expanding into many other trades, such as mechanics, entertainers, meteorologist assistants, pharmacists, service police, and welders. At first they received two-thirds the pay of their male counterparts, although that increased to four-fifths in 1943. They

could be sent "overseas" to England, the United States, or Newfoundland (which was not yet part of Canada). One interesting Canadian family, the Kents, saw their three daughters enlist, as well as their son, who was a pilot. The sisters worked in classified positions, which evidently involved sonar detection, and one of them, at least, assisted in tracking a German submarine up the St. Lawrence River. All four siblings survived the war. James Norton and Dave McKillop, telephone correspondence with author, December 2014; Glad Bryce, *First In, Last Out*, University Women's Club of Toronto, 2010; *On Windswept Heights*, Royal Canadian Air Force, 2013.

4. Tom Hawthorn, "Margaret Fane Rutledge," *Globe and Mail*, January 5, 2005.

5. Render, *No Place for a Lady, The Story of Canadian Women Pilots 1928–1992*, 59.

6. Catherine Atkinson, "Margaret Fane Rutledge," obituary, *Guardian*, January 25, 2005. 155.

Chapter 11: Queen of the Hurricanes: Elsie MacGill

1. Richard I. Bourgeois-Doyle. "Six Decades Later, Elsie MacGill Continues to Inspire," *Society of Women Engineers*, Spring 2011, 28–32.

2. Elizabeth "Elsie" MacGill 1905–1980," Hall of Fame, Canada Technology and Science Museum.

3. Women Aviation Pioneers, *First Canadian 99s*.

4. R. Scott Ward, "Our Great Opportunity," 2008 APTA Presidential Address, *Physical Therapy* 88, no. 10, October 2006.

5. Bourgeois-Doyle, "Six Decades Later, Elsie MacGill Continues to Inspire," 28–32, 30.

6. Ibid., 28–32.

7. Richard (Dick) Bourgeois-Doyle, *Her Daughter, the Engineer: The Life of Elsie Gregory MacGill* (Ottawa: NRC Research Press, 2008).

8. Susan Munroe, "Elsie MacGill," *Canada News*.

9. Bourgeois-Doyle, *Her Daughter, the Engineer*, 32.

10. Bourgeois-Doyle, "Six Decades Later, Elsie MacGill Continues to Inspire."

Chapter 12: Flying Blind: The ATA

1. In the Second World War women in the Soviet Union flew combat missions in three predominantly female regiments. The 588th Air

Regiment and the 587th Bomber Regiment flew bomber missions, the 588th at night. The 586th Fighter Regiment flew air defence missions, and fighter pilot Lily Litvak of that regiment shot down twelve German aircraft and shared the credit for two others. Katya Budanova shot down even more, but the exact number is unknown. Both women were killed in action in 1943. Women in Turkey flew combat missions even earlier — in 1937. In the United States, Eleanor Roosevelt convinced her husband, President F. D. Roosevelt, to create flying roles for women. "This is not a time when women should be patient," she wrote. "We are in a war and we need to fight it with all our ability and every weapon possible. WOMEN PILOTS, in this particular case, are weapons waiting to be used." Thus, the WASP, or Women Airforce Service Pilots, organization was created — 1,100 young female pilots headed by the famous Jacqueline Cochrane. Thirty-eight of them lost their lives ferrying aircraft across the country. "WAI Pioneer Hall of Fame," Women in Aviation International, wai.org.pioneers.

2.   "Women Aviation Pioneers," *First Canadian 99s*. For Vera's story see "Dreams Can Come True," Chapter 7.

3.   Milstead, Vi Warren as told to Arnold Warren, "Flying With the ATA," *CAHS Journal*, Fall 1999.

4.   Hathaway, *Pursuit of a Dream*, 54.

5.   Milstead, Vi Warren as told to Arnold Warren, "Flying With the ATA," *CAHS Journal*, Fall 1999.

6.   Ibid.

7.   "Women Aviation Pioneers," *First Canadian 99s*.

8.   Render, *No Place for a Lady, The Story of Canadian Women Pilots 1928–1992*, 92.

9.   Marilyn Dickson, "Vi Warren to Receive British Badge of Honour," *Northumberland Today*.

10.   Milstead, Vi Warren as told to Arnold Warren. "Flying With the ATA," *CAHS Journal*, Fall 1999.

11.   Hathaway, *Pursuit of a Dream*.

12.   Render, *No Place for a Lady, The Story of Canadian Women Pilots 1928–1992*, 90; "Helen Marcelle Harrison Bristol," Library and Archives Canada, http://epe/lac-bac.gc.ca/100/200/300.

13.   Lou Wise, "I Remember Helen," *CAHS Journal*, Summer 2002, 55.

NEW OPPORTUNITIES
Chapter 13: Into the Woods: Bush Pilots

1.   *Maclean's*, October 15, 1955, 118.
2.   Spring, *Canadian Women Bush Pilots*.
3.   Lorraine R.M. Cooper, *Haiku of Three Worlds*, n.d.
4.   "Lorraine Cooper," obituary, *Calgary Herald*, March 28, 2006.
5.   Eliane/Elianne Roberge Schlachter/Schlageter worked as a secretary for Yukon Southern Air Transport beginning in 1932, but she sometimes served as a co-pilot or spotter for search and rescue and firefighting missions. However, her piloting activities were kept quite secret, although from 1936 on she was part of the Flying Seven in British Columbia. She began flying in 1929. Metcalfe-Chenail, *Polar Winds: A Century of Flying the North*; "Flying Seven," *The History of Metropolitan Vancouver*, n.d.
6.   Metcalfe-Chenail uses the examples of Betty Booth Campbell and her husband Jack Campbell with Spartan Air Services; Dawn and Ron Connelly at the Yukon Flying School in Whitehorse; Molly and Jack Reilly with Peter Bawden Drilling Services in Calgary; and Lorna and Dick de Blicquy, Metcalfe-Chenail. *Polar Winds: A Century of Flying the North*, 114.
7.   Peter Boer, *Bush Pilots* (Edmonton: Folklore Publishing, 2004).
8.   "Vi Flies High at Bash," *Cramahe Now*, October 17, 2009.
9.   "Vi Milstead Warren, C.M.," Canadian Aviation and Space Museum, July 2, 2014.
10. Ibid.
11. "Violet Milstead Warren," *Canadian 99s*.
12. John Chalmers, "41st Annual Induction, Canada's Hall of Fame," May 29, 2014.
13. Sandra Martin, "Lorna de Blicquy," *First Canadian 99s*, March 2009.
14. Ibid.
15. Spring, *Canadian Women Bush Pilots*.
16. "WAI Pioneer Hall of Fame," Women in Aviation International, www.wai.org/pioneers.
17. *Ottawa Journal*, May 11, 1968, 22.
18. Dickson, ed., *Canadian Ninety-Nines 2006 Here and Now*, 86–87; *Ottawa Journal*, May 11, 1968, 22.
19. *Ottawa Journal*, May 11, 1968, 22.
20. Sandra Martin, "Lorna de Blicquy," *First Canadian 99s*, March 28, 2009.

When her daughter was small, it was said that Lorna preferred travelling in buses rather than airplanes. *Ottawa Journal*, May 11, 1968, 22.

21. *Ottawa Journal*, May 11, 1968, 22.

22. June 2, 2004, 3:50 a.m.

23. Designed by Suzanne Wiltshire.

24. Render, *No Place for a Lady*, 214.

25. Spring, *Canadian Women Bush Pilots*.

26. Render, *No Place for a Lady*, 216.

27. "Northern Lights Award — 2013 recipients," Centennial College, www.northernlightsaward.ca/2013_recipient.htm.

28. Spring, *Canadian Women Bush Pilots*.

29. Render, *No Place for a Lady*, 243.

30. Ibid., 244.

31. Spring, *Canadian Women Bush Pilots*.

32. Judy Adamson, telephone and email correspondence with author, 2015.

33. Spring, *Canadian Women Bush Pilots*.

34. Render, *No Place for a Lady*, 206.

35. Ibid.

36. Spring, *Canadian Women Bush Pilots*.

37. NetLetter no. 1289, *Horizons*, June 1978.

38. Render, *No Place for a Lady*, 206.

39. Aviation Leadership Forum, February 19, 1908.

40. Render, *No Place for a Lady*, 299.

41. Spring, *Canadian Women Bush Pilots*.

42. "Air Canada Hires Woman Pilot," *Leader-Post* (Regina), July 14, 1978, 32.

43. Peter Pigott, *Air Canada: The History* (Toronto: Dundurn Publishing, 2014).

44. *Horizons*, June 1978; Judy Evans Cameron, telephone and electronic correspondence with author, 2015.

45. Judy Evans Cameron, telephone and electronic correspondence with author, 2015.

46. Spring, *Canadian Women Bush Pilots*.

47. Lola Reid Allin, electronic communication with author, 2015.

48. Janice Holly Booth, "Adventurista in the Air!" www.janicehollybooth.com/adventurist/your-story/adventurista.

49. Lola Reid Allin, "The Perils and Possibilities of Getting Lost," storyboard; Jan O'Hagan, "Watching the Sunset From Above the Clouds," *Our Town* (Media C Publishing, June 1986).

50. Render, *No Place for a Lady*, 280; "Northern Lights Award — 2012 Award Recipients," Centennial College, www.northernlightsaward.ca/2013_recipient.htm.

51. Mary Ellen Pauli, exhibit speech, Bush Plane Heritage Centre.

52. Ron Grech, "Soaring to Fame," *Daily Press*, February 14, 2011.

53. Ibid.

54. Ibid.

55. Kyle Gennings, "Timmins Pilot Honoured Among Outstanding Women in Aviation," *Daily Press*, November 6, 2012.

56. Anna Pangrazzi, "A Wild Ride," *Helicopter Magazine*, May–June 2013.

Chapter 14: Around the World: Daphne Schiff and Friends

1. Sally Armstrong, "Rare Birds," *Homemakers Magazine*, Summer 1994.

2. Lisa Rundle, "Around the World at 80 Years," *U of T Magazine*, Spring 2013.

3. Ibid.

4. Ibid.

5. "Professor Daphne Schiff," www.20minfo.com/#!search/profile/person?person.

6. Marlene Habib, "Women Take to the Skies to Deliver Medical and Teaching Supplies to Africa," *Canadian Press*, October 10, 2000.

7. "Eleven Distinguished Individuals to Receive Honorary Degrees," *Y-file*, June 9, 2005; Larry Celona and William Neuman, "Foul Play Made Pair Think Plane Went Down," *UFO Updates*, January 13, 2001.

8. Armstrong, "Rare Birds."

9. Toller, "The Sky's The Limit," *Tribune*.

10. Ibid.

11. "Aviators of Distinction — Adele Fogle," *First Canadian 99s*.

12. Toller, "The Sky's The Limit," *Tribune*.

13. "Aviators of Distinction — Adele Fogle," *First Canadian 99s*.

14. Akky Mansikka, "At the Throttle," *Flight Lines* 4, no. 111 (March–April 2012).

15. Adele Fogle, personal correspondence with author, July 6, 2014; Horowitz, "Adele Fogle, Flights of Fancy," *Lifestyles*, Summer 2002, 13.

16. Dickson, ed., *Canadian Ninety-Nines 2006 Here and Now*, 116.

17. Ibid.

18. Habib, "Women Take to the Skies to Deliver Medical and Teaching Supplies to Africa"; Anna Morgan, "Two Women Always Aim for the Sky," *Canadian Jewish News*, October 19, 2011.

19. Marika Kemeny, "Glendon's International Amnesty Club promotes human rights through Air Solidarité," *Y2 File*, May 18, 2005.
20. Dickson, ed., *Canadian Ninety-Nines 2006 Here and Now*, 131.
21. Daphne Schiff, "An Adventure Unlike any Other; Air Solidarité 2001, 'An Uplifting Experience," *Canadian 99s*.
22. Susan McClelland, "Retired Reborn," *Maclean's*, March 3, 2003.
23. Armstrong, "Rare Birds."
24. "Eleven distinguished individuals to receive honorary degrees," *Y-file*, June 9, 2005.

Chapter 15: Spies in the Skies: Operation Skywatch

1. Patricia Bow, "High Flying Honors," *University of Waterloo Communications and Public Affairs*, February 7, 1996.
2. Caroline Lomax, "High Flying Guardians," *Plenty*, no. 25, September 10, 2007.
3. Ralph Wilson, "Female Pilots Fly Pollution Patrols," *Ottawa Citizen*, January 15, 1979; Cameron Smith, "Ontario's Anti-pollution angels," *Toronto Star*, January 6, 1997.
4. Lomax, "High Flying Guardians."
5. Akky Mansikka, email correspondence with author, February 18, 2012.
6. Karen Lloyd, "Becoming Canadian — One Woman's Story," *Canadian Immigrant*, May 25, 2011.
7. Ibid.
8. Ibid.
9. Dickson, ed., *Canadian Ninety-Nines 2006 Here and Now*, 25.
10. Denise Balkissoon, "Local Pilot to Retrace 1910 Loop Over City," *Toronto Star*, July 8, 2010.

Chapter 16: Airplane Gymnastics: The Snowbirds

1. Shirley Render, *No Place for a Lady: The Story of Canadian Women Pilots 1928–1992* (Winnipeg: Portage and Main Press, 1992), 152.
2. "Woman Flies High as New Snowbird's Leader," *Canadian Press*, May 2, 2010.
3. Mike Stroka, *Snowbirds* (Calgary: Fifth House Ltd., 2006); "Death in the Sky: A Snowbird Falls," *CBC News in Review*, February 2005, 47.
4. Darah Hansen, "Top Snowbirds' Pilot Brings Success Story to Vancouver," *Vancouver Sun*, June 25, 2011.
5. Anna Cuthrell, "Woman on Top," n.d.

6.   Ibid.
7.   Jennifer Graham, "She's the Boss: Snowbirds' First Female Pilot to Lead Acrobatic Squad," *Globe and Mail*, May 3, 2010.
8.   *Canadian Press*, May 2, 2010.
9.   "Lieutenant Colonel Maryse Carmichael C.D.," *National Defence News Room*, November 25, 2010.
10. Graham, "She's the Boss: Snowbirds' First Female Pilot to Lead Acrobatic Squad"; Hansen, "Top Snowbirds' pilot brings success story to Vancouver."

Chapter 17: Over the Rainbow: Female Astronauts

1.   Joan Dixon, *Roberta Bondar* (Canmore, Alberta: Altitude Publications, 2006).
2.   Ibid.
3.   Sonia Gueldenpfenning, *Spectacular Women in Space* (Toronto: Second Story Press, 2004).
4.   Dixon, *Roberta Bondar*.
5.   Ibid.
6.   Barbara Bondar with Dr. Roberta Bondar, *On the Shuttle: Eight Days in Space* (Toronto: Greey de Pencier Books, 1993).
7.   Gueldenpfenning, *Spectacular Women in Space*.
8.   Kate Robertson, "To the Moon and Back," *Women of Influence Magazine*, Fall 2011, 31.
9.   Brenda Branswell, "Ambition Accomplished: The Multi-Talented Julie Payette Joins the Roll of Canadian Space Travelers," *Maclean's* 112, no. 20 (May 17, 1999), 57.
10. Chris Gainor, *Canada in Space* (Edmonton: Folklore Publishing, 2006), 164, 210.
11. Robertson, "To the Moon and Back."
12. Guildenpfenning, *Spectacular Women in Space*.
13. Stacey Gibson, *U of T Magazine*, Winter 2009, 40.
14. Ibid.
15. Ibid.
16. Julie Smyth, *Maclean's*, November 12, 2012, 57.

Afterword

1.   "Offensive and Sexist Ads from the Past," Unfinished Man, www.unfinishedman.com/offensive-and-sexist-ads-from-the-past.

2. Johanna Flygare, "A Man's World: The Sex and Gender of Pilots," Lund University School of Aviation, March 14, 2005.

3. "Which Jobs Are Most Compatible?" *eHarmony*, www.eharmony.co.uk/relationship-advice/online-dating-unplug.

4. Not long ago many vocations or professions required women to resign if they married, e.g. stewardesses.

5. See, for example, Elizabeth Gillan Muir, *Petticoats in the Pulpit* (Toronto: The United Church of Canada, 1991).

6. *Canadian Statesman*, February 12, 1857, 1.

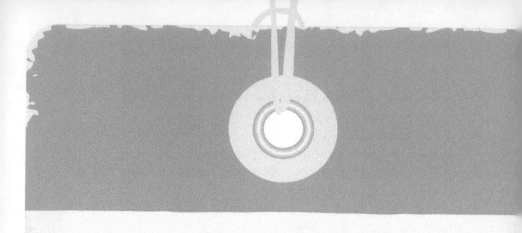

BIBLIOGRAPHY

Periodicals, Newspapers, Pamphlets

"A Moment in Time, Feb. 23, 1909, Silver Dart Makes First Powered Flight in Canada," *Globe and Mail*, February 23, 2012, A2.

"A Nostalgic Album of Old-Time Magazine Art," *Maclean's Magazine*, October 15, 1955, 21–23.

"A Sleeper of Sorts," *The Canadian Aerophilatelist*, December 2002, 6.

Adams, Danielle. "Aviator blazed a trail for other women," *The Globe and Mail*, July 31, 2014, S6.

"Air Canada hires woman pilot," *Leader Post* (Regina, SK), July 14, 1978, 32.

"Air Canada hits anniversary as people's airline," *Ottawa Citizen*, September 2, 1977, 33.

"The Air Hostess," *National Aviation Museum*, 1996.

Aird, Louise. "High Flyers in History: Women in Aviation," *Arts Alive*, March 1998.

"The Almonte Celebration, An Immense Success, Magnificent Balloon Ascension by Miss Thurston," *Almonte Gazette*, September 5, 1897.

"The Amazing Vera Dowling Story," *Max Solbrekken World Missions*, February 18, 2012.

"The American Girl Whom All Europe Is Watching," *World Magazine*, April 11, 1909.

Archer, Christine. "The Sky is not the Limit," *Vanguard*, July 2006.

Armstrong, Sally. "Rarebirds," *Homemakers*, Summer 1994.

"Arrives here by Bird Route from Halifax," *Daily Telegraph* (St. John, N.B.), June 6, 1919, 10.

"Aviation all-stars," *McGill News*, Spring/Summer 2010.

"Aviation Leadership Forum, February 19, 2008, — A Great Success," *Women in Aviation*, no. 2 (March 2008): 4.

Baird, Betty. "treetop airwoman," John McInnes and Emily Hearn. *Sleeping Bags and Flying Machines* (Thomas Nelson & Sons (Canada) Ltd., 1973).

Balkissoon, Denise. "Local pilot to retrace 1910 loop over city," *Toronto Star*, July 8, 2010.

Bourgeois-Doyle. "Six Decades Later, SWE Pioneer Elsie MacGill Continues to Inspire," *SWE*, (Spring 2011): 28–32.

Bow, Patricia. "High Flying Honours," *UW Gazette*, February 7, 1996.

Bradford, R.W. "The Last Flight of La Vigilance," *CAHS Journal*, Winter 1968, 93–98, 108.

Brazier, Kristin. "Lorna de Blicquy," *99 News* (May/June 2009): 30.

Burke, Carl F. "Famous Doctor Jack and Louise Jennings of PEI And Their Associative Prowess," MBE Chapter, *CAHS* 1, no. 19 (June 2006).

Cashman, Tony. "The Katherine Stinson Special," *Canadian Aviation Historical Society* 44, no. 11 (Spring 2006): 4–11.

"Celebrating the Life of Spence McKendry," *Rockliffe Flying Club*, 2009.

"Century of Flight: Wings over Canada," *CBC News in Review* (February 2004): 54ff.

Chase, Sean. "The Extraordinary Escapades of Casey Baldwin," *Daily Observer*, November 8, 2008.

Chase, Sean. "The Incomparable Exploits of Count Jacques de Lesseps," *Daily Observer*, October 14, 2010.

Chouinard, Amy. "Another first for Dr. Wendy Clay," *Canadian Medical Association Journal* 141 (July 15, 1989): 157, 159.

"Class of 2005 encouraged to take the road less travelled," *File*, York University, June 15, 2005.

"Daphne Schiff, Elderly Female Pilot," *University of Toronto Magazine*, Summer 2008.

"de Lesseps, Tauni," Obituary, *New York Times*, March 29, 2001.

"Death in the Sky: A Snowbird Falls," *CBC News in Review* (February 2005): 45ff.

Edgar, Dorothy Leone. "I'm a Flying Housewife," *Chatelaine*, May 1946, 16, 62, 73, 87.

"Eleven distinguished individuals to receive honorary degrees," *File*, York University, June 9, 2005.

Ellis, Frank H., "Pioneer Flying in British Columbia, 1910–1914," *British Columbia Historical Quarterly* 3, no. 4 (October 1939): 4.

Ellis, Frank H. "Katherine Flies the Mail," *Canadian Aviation Historical Society* 12, no. 1 (Spring 1974): 12ff.

Ellis, J.R. "Katherine Stinson in Ontario 1918," *Canadian Aviation Historical Society* 9, no. 1 (Spring 1971): 3.

"The Emancipation of the Cover Girl," *Maclean's Magazine*, October 15, 1955, 28–29, 108.

"Enid McDonald," Obituary, *Guadalajara Reporter*, June 23, 2006.

"Eugene Godardd & The Aurora," *Carnarvon Traders* (Carnarvon Wales), n.d.

Fitzgerald, P. "A Career in Aviation," *CAHS Journal* (Spring 2003): 4–17.

Flygare, Johanna. "A man's world, The sex and gender of pilots," Lund University School of Aviation, March 14, 2005.

Forsythe, Bruce. "Vaughan Perspectives in Canadian Military History," August 30, 2008.

Fortier, Renald. "The Balloon Era," *Canada Museum*, 2004.

Franklin, Jasmine. "Former Fighter Pilot says women should aim high," *Edmonton Sun*, March 14, 2011.

Fuller, George. "The First Canadian Woman to Fly," *CAHS Journal* 7 (1969).

Gennings, Kyle. "Timmins pilot honoured among outstanding women in aviation," *Daily Press*, November 6, 2012.

"Gets Pilot's Licence," *Evening Telegram*, March 15, 1928, 27.

Gibson, Stacey. "Ms. Universe," *University of Toronto Magazine*, Winter 2009, 38–43.

"Glendon's Amnesty International Club Promotes Human Rights Through 'Air Solidarité,'" *File*, York University, May 18, 2005.

Graham, Jennifer, "She's the boss, Snowbirds' first female pilot to lead aerobatic squad," *Globe and Mail*, May 3, 2010.

Graham, Stuart with R.E. Graham. "The beginning of Bush Flying," *CAHS Journal* (Winter 2002): 124–128, 131–137.

"Grand Balloon Ascension at McFarlane's Grove in 1879," *Almonte Gazette*, reprinted July 30, 1970.

Grant, Doug. "Aeronaut Prof. Squires [*sic*] Provides an Exciting Day in Brockville — 1974," *Handbook of Brockville History*, n.d.

Gray, Charlotte. "Roberta Bondar," *Chatelaine*, June 1984, 50, 74, 76, 80, 88.

Grech, Ron. "Soaring to Fame," *Daily Press*, February 14, 2011.

"Guelph Airpark in limbo after flight school's recent closure," *Guelph Mercury*, April 16, 2011.

Habib, Marlene. "Women take to the skies to deliver medical and teaching supplies to Africa," *Canadian Press*, October 10, 2000.

"Hamilton Girl Earns Right to Pilot Airplane," *Hamilton Spectator*, March 14, 1928.

"Hamilton Girl First Canadian Woman Pilot," *Toronto Daily Star*, March 15, 1928, 1.

Hansen, D.A. "Top Snowbirds pilot brings success story to Vancouver," *Vancouver Sun*, January 25, 2011.

Hawthorne, Tom. Obituary, "Margaret Fane Rutledge," *Globe and Mail*, January 5, 2005, 57.

"High-flying honors," *University of Waterloo Magazine*, July 2, 1996.

Hitchins, F.H. "Ballooning was safer in Canada," *CAHS Journal* (Summer 1947): 45f.

Hoh, John L., Jr. "Air Canada: Swept into the Sky With Maple Leaves," *Airlines*, September 12, 2005.

Holden, Lt. Col. Ron. "I Remember Violet Milstead Warren," *Globe and Mail*, August 2, 2014, S9.

Horowitz, Dan. "Adele Fogle, Flights of Fancy," *Lifestyles*, Summer 2002, 13, 15.

Janus, Andrea. "Female Pilot in 'shock' after passenger leaves sexist note," *CTV News*, March 4, 2014.

Jensen, Sandra. "Learning to Fly," *Stories of the Stuart Graham Archives*, Concordia University, 2010.

Johnson, Brian. "'It's been a magical ride," *Kingston Whig Standard*, April 29, 2010.

"Joyce Bond, Aviatrix from Regina," *Canadian Aviation Historical Journal* 5, no. 3 (Fall 1967).

Kelly, Evelyn. "She's Boss Stewardess of the Atlantic Hop," *Chatelaine*, July 1947, 10.

Kemeney, Marika. "Glendon's International Amnesty Club promotes human rights through Air Solidarité," Y2 file, May 18, 2005.

Kwasnik, Andrea. "History Maker," *Wings*, 2013.

"*The Lafayette Negative Archive*," New York City.

"Last Month's Meeting," *Flypast*, Canadian Aviation Historical Society, October 27, 2007.

Léger, Richard J. "The Air Hostess," *National Aviation Museum*, 1996.

Lienhard, John H. "No. 1251: Katherine Stinson," *Engines of Our Ingenuity*, n.d.

Lloyd, Karen. "Becoming Canadian, one woman's story," *Canadian Immigrant*, May 25, 2011.

Lomax, Carolyn. "High Flying Guardians," *Plenty Magazine*, September 7, 2010.

"Lorraine Cooper," Obituary, *McInnis & Holloway Funeral Homes*, March 2006.

"Lucile Garner was Canada's first airline stewardess," Obituary, *Globe and Mail*, March 28, 2013.

Martin, Sandra. "Lorna de Blicquy, 77: Trailblazing Pilot," *Globe and Mail*, March 28, 2009.

Masterman, Bruce. "Bush Pilots," *West*, Spring 2005.

McClelland, Susan. "Retired Reborn," *Maclean's*, March 3, 2003.

McKendry, Felicity. "Letter to the Editor," *Kingston Whig-Standard*, September 27, 2012.

Meszaros, Capt. Nicole and 2 Lt. Alex Whittaker. "A Canadian Icon Celebrates a Milestone," *National Defence Press Release*, October 29, 2010.

Mick, Haley, "Six decades later, Olympic hero still inspires generations," *Globe and Mail*, October 2, 2012, S7.

"The Mile-High Club," *Walrus*, June 2010, 21.

Mills, Albert J., and Jean Helm Mills. "Masculinity and the Making of Trans-Canada Air Lines, 1938–1940: A Feminist Poststructuralist Account," *Canadian Journal of Administrative Sciences*, 2006.

Milstead, Vi Warren, as told to Arnold Warren. "Flying With the ATA," *CAHS Journal*, Fall 1999, 84–93, 108–112.

Morgan, Anna. "Two Women Always Aim For the Sky," *The Canadian Jewish News*, February 11, 2005.

Morris, Hobie. "Central New York's First Aerial Heroine: Balloonist Nellie Thurston Squire," *Musings of a Simple Country Man*, n.d.

Munroe, Susan. "Elsie MacGill," Faculty of Interior Design, RCC, *Institute of Technology*, n.d.

"Nellie Thurston's Account of the Ascension at Watertown, N.Y.," *Almonte Gazette*, August 1, 1879.

O'Hagan, Jan. "Watching the Sunset From Above the Clouds," *Our Town*, Media C Publishing, June 1986.

Pangrazzi, Anna. "A Wild Ride," *Helicopter Magazine*, May–June 2013.

"Plans to fly from Halifax to Quebec Without a Stop," *Daily Telegraph* (St. John, N.B.), June 18, 1919, 5.

Potter, Kelly. "Trailblazing the Airways," *Wings*, May 4, 2009.

"Quebec's first lift-off was in a balloon named Canada," *Quebec Heritage News* 2, no. 11 (July 2004), 7.

"Queen of the Blades," Between Ourselves, *TCA Employee Newsletter*, February 1947.

Quimby, Harriet. "A Woman's Exciting Ride in a Racing Motor-Car," *Leslie's Weekly*, October 4, 1900.

Reimer, Isobel. "Women on the Go — Rosella Bjornson," *Chatelaine*, May 1974, 12.

Ricketts, Bruce. "Madge Graham … Canada's First Lady of Flight," *Mysteries of Canada*, 1998–2011.

Robertson, Kate. "To the Moon and Back," *Women of Influence Magazine*, Fall 2011, 30.

Robertson, Peter. "Joyce Bond, Aviatrix from Regina," *Canadian Aviation Historical Society* 5, no. 3 (Fall 1967): 68.

Royal Canadian Air Force. "Attractive Careers," *Maclean's Magazine*, October 15, 1955, 118.

Rundle, Lisa. "Around the World in 80 Years," *U of T Magazine*, Summer 2005.

Rungeling, Dorothy. "Felicity McKendry — She had a dream," *Kingston Flying Club*, October 20, 2002.

Russo, Carolyn. "Women and Flight," *National Air and Space Museum*.

Ryell, Nora. "She was Canada's first airline stewardess," *Globe and Mail*, March 29, 2013, 8.

Saner, Emine. "Female pilots: a slow take-off," *Guardian*, January 13, 2014.

Schiff, Daphne. "Flying in Africa an Uplifting Experience," Toronto, 2005.

Schliesmann, Paul. "Helping Dreams Take Flight," *Kingston Whig Standard*, June 5, 2009.

"She Taught Pilot to Fly," *Sarasota Journal*, February 4, 1972, 3.

Sissons, Crystal. "Elsie Gregory MacGill: Engineer, Feminist and Advocate for Social Change," *Atlantis* 34, no. 1 (2009): 48.

"Skating Queen Likes Flying," *Canadian Aviation*, April 1947, 95.

Smith, Cameron. "Ontario's anti-pollution angels," *Toronto Star*, January 6, 1997.

Smyth, Julie. "Julie Payette," *Maclean's*, November 12, 2012, 57.

"Snowbird Formations are Gender-Blind," *Maclean's*, May 17, 2010, 25.

"Spitfires in the Rhododendrons," *CBC Digest* Archives, August 2003.

Stanonis, Anthony J. "Book review of Femininity in Flight by K.M. Barry," *Canadian Journal of History*, 1997.

"Stewardess tells odd tale," *Ottawa Evening Citizen*, January 20, 1948.

"Successful flights at Montreal Meet," *Knoxville Journal*, June 29, 1910.

"Tall girls and tiny aircraft," *Star-Phoenix*, February 3, 1971, 23. Republished by *Canadian Press*.

Taylor, Paul. "End of the Shuttle era," *Globe and Mail*, July 8, 2011, A10f.

"TCA's First Stewardess, Lucile Garner Gant," *NetLetter for Air Canada's Retirees*, no. 1126, June 26, 2010.

"TCA Girls Hold Reunion," *Horizons*, July 1973, *NetLetter* #1239.Theobold, Claire. "Despite long, proud history of participation on the tarmac and in the sky, female pilots are still a rarity," *Edmonton Examiner*, March 17, 2014.

Tingley, Ken. "Historical Impact Assessment, A Thematic Overview Narrative for City Centre Airport," June 2009, Report 2009DCM032.

Toller, Carol. "The Sky's The Limit," *Chicago Tribune*, November 5, 1995.

Traviss, Wendy. "On the Wings of a Nightingale, TCA's Flying Ladies," *Canadian Aviation Historical Society*, Spring 1986.

"Vi flies high at bash," *Cramahe Now*, October 17, 2009.

Vollick, Eileen, M. "How I became Canada's first licensed woman pilot," *Owen Sound Sun Times*, August 6, 2008.

Ward, Scott. "Our Great Opportunity," *Physical Therapy* 88, no. 10 (October 2008).

"Was Quebecer Stuart Graham the first bush pilot?" *Quebec Heritage News* 2, no. 11 (July 2004): 11.

Watson, Patrick. "Marion Alice Orr," *Canadians, Biographies of a Nation* 3, 2002.

[Wedding notices] *Times*, January 26, 1911, 11a.

Wedemeyer, D. "Katherine Stinson Otero, 86, dies: Pioneer Aviator and Stunt Flier," *New York Times*, July 11, 1977, 28.

"Will stop here on flight from Halifax to Three Rivers, Quebec," *Daily Telegraph* (St. John, N.B.), June 5, 1919.

Wilson, Ralph. "Female pilots fly pollution patrols," *Ottawa Citizen*, January 15, 1979.

Wise, Lou. "I Remember Helen," *CAHS Journal*, Summer 2001, 52–55.

"Wit and Wisdom," *Maclean's Magazine*, October 15, 1955, 101.

"Woman Flies High as New Snowbird's Leader," *Canadian* Press, May 2, 2010.

"Women in Aviation," Canadian Bushplane Heritage Centre special exhibition text, 2011.

Zilm, Glennis. "'I Taught the Pilot to Fly' She Said to Skeptic," *The Blade: Toledo, Ohio*, February 27, 1972, 4.

Books

Baker, Trudy, and Rachel Jones with Donald Bain. *Coffee, Tea or Me?* (New York: Penguin Books, 2003).

Barry, Kathleen M. *Femininity in Flight: A History of Flight Attendants* (Duke University Press, 2007).

Boer, Peter. *Bush Pilots* (Edmonton: Folklore Publishing, 2004).

Bondar, Barbara, with Dr. Roberta Bondar. *On the Shuttle — Eight Days in Space* (Toronto: Greey de Pencier Books, 1993).

Bryce, Glad. *First In, Last Out* (Toronto: University Women's Club, 2010).

Davison, Rowena, with Reuben Davison. *An Edwardian Housewife's Companion* (Newberry Park California: Haynes Publishing, 2009).

Dickson, Marilyn, ed., *Canadian Ninety-Nines 2006 Here and Now* (Toronto: Margo McCutcheon, 2007).

Dixon, Joan. *Roberta Bondar: The Exceptional Achievements of Canada's First Woman Astronaut* (Toronto: James Lorimer Limited, 2006).

Dublin, Anne. *June Callwood, A Life of Action* (Toronto: Second Story Press, 2007).

Dundas, Barbara. *A History of Women in the Canadian Military* (Montreal: Art Global, 2000).

Fleming, R. B. *The Railway King of Canada* (University of British Columbia Press, 2013).

Forster, Merna. *100 Canadian Heroines* (Toronto: Dundurn Press, 2004).

Gainor, Chris. *Canada in Space* (Folklore Publishing, 2006).

Gueldenpfenning, Sonia. "Roberta Bondar" and "Julie Payette" in *Spectacular Women in Space*, (Toronto: Second Story Press, 2004).

Hathaway, Warren. *Pursuit of a Dream, The story of pilot Vera Strodl Dowling* (Warren Hathaway, 2012).

Jackson, Donald Dale. *The Aeronauts* (Alexandria, Virginia: Time-Life Books, 1980).

Lovett-Reid, Patricia. *Get Real — 26 Canadian Women Share the Secret to Authentic Success*, (Toronto: Key Porter Books, 2007).

McKendry, Felicity. *I Grew Too Tall* (Ottawa: Felicity McKendry, 2012).

Metcalfe-Chenail, Danielle. *Polar Winds — A Century of Flying the North* (Toronto: Dundurn Press, 2014).

Morrow, Don and Kevin B. Wamsley. *Sport in Canada, A History* (Don Mills, Ontario: Oxford University Press, 2010).

Newby, Jill. *The Sky's the Limit* (Vancouver: Mitchell Press, 1986).

Payne, Stephen ed. *Canadian Wings* (Rockliffe: Canadian Aviation Museum, 2006).

Prentice, Alison et al. *Canadian Women, A History* (Toronto: Harcourt Brace Jovanovitch, 1988).

Render, Shirley. *No Place for a Lady, The Story of Canadian Women Pilots 1928–1992*, (Winnipeg: Portage and Main Press, 1992).

Royal Canadian Air Force. *On Windswept Heights II* (Royal Canadian Air Force, 2013).

Smyth, Ross. *The Lindbergh of Canada: The Erroll Boyd Story* (Renfrew, Ontario: General Store Publishing House, 1997).

Speaking of Flying (San Clemente, California: Aviation Speakers Bureau, 2000).

Spring, Joyce. *Canadian Women Bush Pilots* (Toronto: Natural Heritage Books, 2006).

Spring, Joyce. *Daring Lady Flyers* (Lawrencetown Beach, N.S.: Potterfield Press, 1994).

Sroka, Mike. *Snowbirds* (Calgary: Fifth House Ltd., 2006).

The Ninety-Nines — Yesterday — Today — Tomorrow (Paducah, Kentucky: Turner Publishing Co., 1996).

Wojna, Lisa. *Great Canadian Women* (Folklore Publishing, 2005).

Websites

"Adele Fogle," http://womenaviators.org/wiki/index.php?title=Adele_Fogle.

Allen, Shirley, "The Ninety-Nines," www.canadian99s.org/articles/rarebirds.htm.

Allin, Lola Reid, "The Perils and Possibilities of Getting Lost," www.vergemagazine.com/storyboard.

"Danish WW2 Pilots: Strodl, Vera Elsie (1918–)," www.danisjww2pilots.dk/profiles.php?alpha=s&id=164.

"Elsie MacGill," http://womynherstory.blogspot.com/2009/10/elsie-macgill.html.

"Femininity in Flight, A History of Flight," www.c.gc.eng/mediaroom/infosheets-women_pilots-1351.htm.

Find a Grave, www.findagrave.com.

"Flying in Africa: an Uplifting Experience," http://sf260w.com/AirSoli.html.

"Flying Seven," www.Vancouverhistory.ca/archives_flying_seven_htm.

"Flying to help the needy in Africa," www.yorku.ca/yfile/archive/index.asp?Article=3929.

"Great Aviation Quotes," www.skygod.com/quotes/highflight.html.

Grech, Ron, "Soaring to Fame," www.thedailypress.ca/ArticleDisplay.aspx?e=2977791&archive=true.

Lane, Robert K., "Re-Enactment of Western Canada's First Air Mile Flight," www.bnaps.org/ore/Lane-Stinson/Lane-Stinson.htm.

Lloyd, Karen, "Becoming Canadian — one woman's story," http://canadianimmigrant.ca/immigrant-stories/becoming-canadian-%E2%80%94-one-woman's.

Mansikka, Akky, "Flying the Gold Cup Rally 2003," www.canadian99s.org/ECAN/gcrally/2003rally.htm.

"Margo McCutcheon," www.northernlightsaward.ca/judge.html.

"Offensive and Sexist Ads from the Past," www.unfinishedman.com/offensive-and-sexist-ads-from-the-past.

"The Remarkable Harriet Quimby," www.todayifoundout.com.

Ricketts, Bruce, "Canada Car & Foundry," www.mysteriesofcanada.com/Canada/Canada-car/ccf_part_2_elsie_macgill.htm.

"Roberta Bondar" and "Julie Payette," www.collectionscanada.gc.ca/women.

"Stewardesses of the 1950s," http://epe.lac-bac.gc.ca/100/200/301/ic/can_digital_collections/high_flyers/steward.htm.

Spillen, Alan K., "Katherine Stinson in Canada," http://earlyaviators.com/estinka2.htm.

"TCA's first stewardess," June 15, 2010, www.canflyer.com/canflyer/viewtopic.php?f=1&t=5519.

"Women in Aviation International," www.wai.org/pioneers/100women.

"Which jobs are the most compatible?" www.eharmony.co.uk/relationship-advice/online-dating-unplug.

"You've Got Mail … an Aviatrix in Alberta," www.vintagewings.ca/VintageNews/Stories/tabid/116/articleType/.

INDEX